STATE OF THE
WORLD'S FORESTS
2009

FOOD AND AGRICULTURE ORGANIZATION OF THE UNITED NATIONS

Rome, 2009

Produced by the
Electronic Publishing Policy and Support Branch
Communication Division
FAO

ISBN 978-92-5-106057-5

Contents

Foreword

State of the World's Forests, published biennially, provides a global view of major developments affecting forests. The theme for the 2009 edition is "Society, forests and forestry: adapting for the future".

The 2007 issue reviewed "Progress towards sustainable forest management" with an emphasis on the "supply side", in particular forest resources. *State of the World's Forests 2009* places more emphasis on the "demand side": What will be the impact on forests of future increases in global population, economic development and globalization? Is the explosion in global trade having positive or negative effects on the world's forests? Will the forest sector continue to have a major role in providing livelihoods for rural communities?

This eighth edition looks forward. Part 1 summarizes the outlook for forests and forestry in each region of the world. FAO periodically carries out regional forest sector outlook studies in collaboration with countries and organizations in each region. The results of studies for all regions are summarized and presented here for the first time in a single publication. A main pattern that emerges is a strong correlation between economic development and the state of forests. Countries that are undergoing rapid economic growth tend to struggle with immense pressures on their forests. In contrast, regions that have already achieved a high level of economic development are usually able to stabilize or increase their forest area. However, the factors affecting forests are numerous and complex, making it difficult to draw simple conclusions or to make reliable projections.

Part 2 considers how forestry will have to adapt for the future. It begins with a global outlook for wood products demand to 2030, noting changing patterns in production, consumption and trade. Next, a chapter on environmental services of forests probes the various market and non-market mechanisms evolving to help forests and trees fulfil their environmental service functions of land, water

and biodiversity protection, carbon storage and others. A look at progress in institutional adaptation notes that many forestry institutions are having difficulty in adapting to rapid changes in communications, globalization and society's expectations. Those institutions that are willing and able to adapt are more likely to be successful in the future. Finally, Part 2 examines developments in science and technology, which will continue to have an enormous impact on the future of forests and forestry. Imagine a world in which trees are a major source of fuel for cars, replacing oil. Only a few years ago, this seemed like fantasy, but today the possibility must be seriously considered.

As this edition was going to press in late 2008, the global economy had spun into sharp decline, sparked by the contraction of the housing sector and the subprime mortgage crisis in the United States of America. Almost all countries have been affected by this downturn. While these events occurred too late to be incorporated in the main text of *State of the World's Forests 2009*, a postscript has been added addressing the observed and likely impacts of the economic crisis on forests and forestry. It notes that while there is considerable uncertainty about how the situation will unfold in the coming years, the crisis could also provide an opportunity to chart a new path for the development of the forest sector.

State of the World's Forests 2009 has two main goals. As with past editions, it is intended to serve as a source of information to support policy and research related to forests. In addition, I hope that it helps to stimulate creative thinking and debate about the future of the world's forests.

Jan Heino
Assistant Director-General
FAO Forestry Department

Acknowledgements

The preparation of *State of the World's Forests 2009* was coordinated by C.T.S. Nair. Special thanks go to A. Perlis, who edited the publication, and to R. Rutt, who provided research support.

The following FAO staff wrote or reviewed sections of the report or assisted with tables, maps, graphics or other information: M. Achouri, G. Allard, B. Amado, S. Appanah, J.L. Blanchez, M. Boscolo, S. Braatz, A. Branthomme, J. Broadhead, C. Brown, J. Carle, C. Carneiro, F. Castañeda, M. Chihambakwe, R. Czudek, P. Durst, C. Eckelmann, T. Etherington, P. Evans, V. Ferreira, B. Foday, M. Gauthier, A. Gerrand, S. Grouwels, J. Heino, S. Hetsch, T. Hofer, P. Holmgren, A. Inoguchi, O. Jonsson, R. Jonsson, F. Kafeero, W. Killmann, D. Kneeland, P. Koné, M. Laverdiere, A. Lebedys, M. Lobovikov, Q. Ma, L. Marklund, R.M. Martin, M. Morell, E. Muller, F. Padovani, M. Paveri, E. Pepke, J.A. Prado, C. Prins, D. Reeb, D. Rugabira, O. Serrano, O. Souvannavong, R. Suzuki, T. Vahanen, P. Vantomme, A. Whiteman, M.L. Wilkie and J. Zapata-Andia.

FAO specifically thanks, for contributions and reviews: L. Alden Wily, D. Baskaran Krishnapillay, S. Boucher, M. Boyland, J. Campbell, J. Cinq-Mars, A. Kaudia, R. Keenan, L. Langner, J. Maini, E. Mansur, P. O'Neill, J. Parrotta, R. Persson, F. Raga Castellanos, M.A. Razak, R. Sedjo, J. Severino Romo, H.C. Sim and E. Sirin.

FAO also acknowledges members of the Collaborative Partnership on Forests (CPF) for their contributions presented in Part 2, specifically the Convention on Biological Diversity (CBD), the Center for International Forestry Research (CIFOR), the International Tropical Timber Organization (ITTO), the International Union of Forest Research Organizations (IUFRO), the Global Mechanism of the United Nations Convention to Combat Desertification (UNCCD), the United Nations Environment Programme (UNEP), the United Nations Forum on Forests (UNFF) and the United Nations Framework Convention on Climate Change (UNFCCC).

The staff of the FAO Electronic Publishing Policy and Support Branch provided editorial and production support.

Abbreviations and acronyms

CAMPFIRE Communal Areas Management Programme for Indigenous Resources
CBD Convention on Biological Diversity
CDM Clean Development Mechanism
CIFOR Center for International Forestry Research
CIS Commonwealth of Independent States
CPF Collaborative Partnership on Forests
ECOWAS Economic Community of West African States
EFSOS European Forest Sector Outlook Study
FECOFUN Federation of Community Forest Users Nepal
FLEG forest law enforcement and governance
FSC Forest Stewardship Council
GCC Gulf Cooperation Council
GDP gross domestic product
GIS geographic information system
ICT information and communication technology
IEA International Energy Agency
IIED International Institute for Environment and Development
IPCC Intergovernmental Panel on Climate Change
ITTO International Tropical Timber Organization
IUCN International Union for Conservation of Nature
IUFRO International Union of Forest Research Organizations
LULUCF land use, land-use change and forestry
MCPFE Ministerial Conference on the Protection of Forests in Europe
MTOE million tonnes oil equivalent
NAFTA North American Free Trade Agreement
NGO non-governmental organization

NLBI Non-Legally Binding Instrument on All Types of Forests
NWFP non-wood forest product
PEEN Pan-European Ecological Network
PEFC Programme for the Endorsement of Forest Certification Schemes
PES payments for environmental services
R&D research and development
REDD reducing emissions from deforestation and forest degradation
REIT real estate investment trust
SADC Southern African Development Community
SWF sovereign wealth fund
TIMO timber investment management organization
UN United Nations
UNCCD United Nations Convention to Combat Desertification
UNCED United Nations Conference on Environment and Development
UNECE United Nations Economic Commission for Europe
UNEP United Nations Environment Programme
UNESCAP United Nations Economic and Social Commission for Asia and the Pacific
UNESCO United Nations Educational, Scientific and Cultural Organization
UNFCCC United Nations Framework Convention on Climate Change
UNFF United Nations Forum on Forests
WRI World Resources Institute
WRME wood raw-material equivalent

Summary

This eighth biennial issue of *State of the World's Forests* considers the unfolding future of forests and forestry at the subregional, regional and global levels. Based on the most recent of FAO's periodic forest sector outlook studies, it examines the impacts that external factors such as demographic, economic, institutional and technological changes may have on forests. With globalization and improved communications, the regional scenarios will be increasingly interlinked. However, some countries and regions are better prepared to face the upcoming challenges and take advantage of emerging opportunities, while others still lack essential institutional, legal and economic conditions to manage their forest resources in a sustainable manner.

PART 1: REGIONAL OUTLOOK

Africa

The forest situation in Africa presents enormous challenges, reflecting the larger constraints of low income, weak policies and inadequately developed institutions. The growing population and rising prices of food and energy will exacerbate the situation, especially as increased investments in infrastructure open up new areas. Progress in implementing sustainable forest management is expected to be slow, with forest loss likely to continue at current rates.

The forest outlook will depend greatly on political and institutional developments – on improved efficiency and accountability in the public sector; greater inclusiveness, competitiveness and transparency in market institutions; and an informal sector that provides increased livelihood opportunities for the poor. Focusing on products and services required locally and globally and strengthening local institutions can be important ways of addressing forest resource depletion. Such efforts should build on local knowledge and experience of sustainable resource management integrating agriculture, animal husbandry and forestry.

Asia and the Pacific

Considering the great diversity in Asia and the Pacific, several scenarios are expected to unfold. While forest area will stabilize and increase in most of the developed countries and some of the emerging economies, the low- and middle-income forest-rich countries will witness continuing declines as a result of expansion of agriculture, including the production of biofuel feedstock.

Demand for wood and wood products will continue to increase in line with the growth in population and income. Growth in the demand for primary commodities owing to rapid industrialization of emerging economies is likely to result in forest conversion in other countries within and outside the region. While the region is a leader in planted forest development, it will continue to depend on wood from other regions, as land and water constraints will limit the scope for self-sufficiency in wood and wood products. The demand for forest environmental services will increase as incomes rise, and conservation involving local communities is likely to receive greater emphasis.

Europe

Forest resources in Europe are expected to continue to expand in view of declining land dependence, increasing income, concern for protection of the environment and well-developed policy and institutional frameworks. The provision of environmental services will remain a primary concern, especially in Western Europe, and rules and regulations will make wood production less competitive than in other regions.

Forest management will continue to serve a wide variety of purposes. Economic viability is likely to remain a challenge, especially for small-scale forest owners, but the increased demand for woodfuel could change this. While the forest industry, especially in Western Europe, may continue to lose competitiveness with other regions in labour-intensive segments, it is likely to retain leadership in the production of technologically advanced products. Within the region, the differences in forestry between Eastern and Western Europe are likely to diminish as Eastern Europe catches up economically.

Latin America and the Caribbean

Forests and forestry in Latin America and the Caribbean will be influenced by the pace of economic diversification and changes in land dependence. In Central America and the Caribbean, where population densities are high, increasing urbanization will cause a shift away from agriculture, forest clearance will decline and some cleared areas will revert to forest. However, in South America, the pace of deforestation is unlikely to decline in the near future despite low population density. High food and fuel prices will favour continued forest clearance for production of livestock and agricultural crops for food, feed and biofuel to meet the global demand. Sustainable forest management will continue to be a challenge in a number of countries where land tenure is poorly defined.

Latin America and the Caribbean has considerable opportunities to benefit from the growing demand for global public goods provided by forests, especially carbon sequestration and storage, but realizing the potential will require substantial improvements in policy and institutional frameworks. Planted forests will increase, promoted by private investments and continuing global demand for wood and wood products from Asia. However, it is unlikely that the increased planting rate will be sufficient to offset continuing deforestation.

North America

The near future of North American forestry will depend on how quickly the region reverses the recent economic downturn and its impact on the demand for wood and wood products, especially in the United States of America. The forest sector will also need to address challenges of climate change, including increasing frequency and severity of forest fires and damage by invasive pest species. Wood will be increasingly demanded as a source of energy, especially if cellulosic biofuel production becomes commercially viable; this development would likely result in much larger investments in planted forests.

Canada and the United States of America will continue to have fairly stable forest areas, although the divestment of woodlands owned by large forest companies could affect their management. In Mexico, changes in the deforestation rate will depend on the pace of transition from an agrarian to an industrial economy and reduced dependence on land as a source of income and employment. While the economic viability of the forest industry may fluctuate and even decline, the provision of environmental services will continue to gain in importance, driven by public interest.

Western and Central Asia

The outlook for forests and forestry in Western and Central Asia is mixed. Income growth and urbanization suggest that the forest situation will improve or remain stable in some countries, but the picture is less promising for a number of low-income agriculture-dependent countries. Forest degradation will persist in countries that are relatively well off but have weak institutions. Overall, the forest sector is given low priority in public investments.

Adverse growing conditions limit the prospects for commercial wood production. Rapidly increasing incomes and high population growth rates suggest that the region will continue to depend on imports to meet demand for most wood products. Provision of environmental services will remain the main justification for forestry, especially arresting land degradation and desertification, protecting watersheds and improving the urban environment. Institution building, particularly at the local level, is needed in order to facilitate an integrated approach to resource management.

PART 2: ADAPTING FOR THE FUTURE
Global wood products demand

The income that owners derive from managing forests to meet the demand for goods and services is a primary determinant of investment in forest management. Demographic changes, economic growth, regional economic shifts and environmental and energy policies will be decisive in the long-term global demand for wood products.

Production and consumption of key wood products and wood energy are expected to rise from the present to 2030, largely following historical trends. The most dramatic change will be the rapid increase in the use of wood as a source of energy, particularly in Europe, as a result of policies promoting greater use of renewable energy. The highest growth rates will continue to be in Asia, which will be the major producer and consumer of wood-based panels and paper and paperboard (although per capita consumption will remain higher in Europe and North America). Industrial roundwood production in Asia will be far short of consumption, increasing the dependence on imports.

The potential for large-scale commercial production of cellulosic biofuel will have unprecedented impacts on the forest sector. Increasing transport costs will also influence the demand for wood products. These factors and others, including changes in exchange rates, will influence the competitiveness of the forest sector and affect production and consumption of most forest products.

Industrial roundwood will be increasingly likely to come from planted forests in the future. This continuing shift presents interesting opportunities and challenges for forest management.

Meeting the demand for environmental services of forests

Growth in income coupled with greater awareness will increase the demand for environmental services provided by forests such as clean air and water, mitigation of climate change and unspoilt landscapes. While income growth also enhances the ability of society to meet the costs of environmental protection, economic growth is often accompanied by increased environmental impacts. In particular, countries with rapidly growing economies often go through a period when forest resources are exploited and their environmental services decline accordingly. Maintaining forest environmental services requires striking a balance between the production of goods and the provision of services.

Regulatory approaches for helping to secure forests' ability to meet the demand for environmental services include protected areas, instruments for sustainable forest management and green public procurement policies.

Market approaches include certification, carbon markets and payments for environmental services (PES). Third-party certification of forests will continue to spread, although obtaining a premium to cover implementation costs remains a challenge. Systems for providing appropriate payments to forest owners as a means of supporting forest conservation are receiving considerable attention; they have long existed for recreational services and are now being adopted for watershed protection, biodiversity conservation and carbon sequestration. Such schemes are expected to increase in number; stable institutional and legal frameworks are a prerequisite for their success.

Ongoing discussions about including options for reducing emissions from deforestation and forest degradation (REDD) in global climate change negotiations have raised many hopes. However, providing incentives to desist from deforestation involves complex policy, institutional and ethical issues that must be considered.

Changing institutions

The shifting balance among forest sector institutions – public agencies, the private sector, civil-society organizations, the informal sector and international organizations – will play an important part in society's adaptation to social, economic and environmental change. With the emergence of new players, the institutional landscape has become more complex. In general (although not in all countries), the playing field is becoming more level, partly as a result of new information and communication technologies. Much-needed pluralism provides new opportunities for small and medium enterprises and community organizations. If government forestry agencies that have historically dominated the scene fail to adapt to change, they could fade into irrelevance.

With the increasing pace of globalization, new players such as timber investment management organizations (TIMOs), real estate investment trusts (REITs), sovereign wealth funds and carbon-trading institutions could alter the global institutional map. Institutions will face tremendous pressure to balance fragmentation and to consolidate efforts.

Developments in forest science and technology

Visualizing the future of forest science and technology is difficult given the rapid pace of change. Innovation has significantly improved the capacity of the forest sector to meet the changing demands of society and will continue to do so. However, many developing countries have little or no credible science capacity, and this lack hinders their long-term development. Even in many developed countries, forest science and technology capacities have eroded.

However, research continues to break new ground in all areas of forestry, from production, harvesting and processing to wood energy and the provision of environmental services. Relatively new fields such as biotechnology, nanotechnology and information and communication technologies contribute to these developments. The value of indigenous knowledge is increasingly being recognized.

The growth of commercially driven private-sector research and the declining capacity of public-sector research raise a number of issues. Vast populations that cannot afford to pay for improved technologies are often excluded from the benefits of private-sector research. This accentuates disparities in access to knowledge, with consequences for income and living standards.

More concerted efforts are needed to address imbalances and deficiencies in scientific and technological capacity. Challenges include reducing barriers to the flow of technologies among and within countries, ensuring that social and environmental issues are mainstreamed, and transcending traditional sectoral boundaries to take advantage of scientific and technological developments outside the forest sector.

Postscript: challenges and opportunities in turbulent times

As *State of the World's Forests 2009* goes to press (late 2008), the world is experiencing a steep economic decline. The contraction of the housing sector and the subprime mortgage crisis in the United States of America have severely affected financial markets, triggering an economic slowdown involving almost all countries and transforming previously upbeat economic forecasts.

What impacts will these changes have on the forest sector? The collapse of the housing sector has reduced the demand for a wide array of wood and wood products, leading to mill closures and unemployment. New investments are slowing as a result, affecting all wood industries.

The demand for environmental services has also changed as a result of reduced ability and willingness to pay for such services. Carbon prices have remained highly volatile. Future climate change arrangements may face challenges as countries give priority to tackling the economic crisis.

Potential negative impacts on forest resources could include reduced investment in sustainable forest management and a rise in illegal logging as the decline

in the formal economic sector opens opportunities for expansion of the informal sector. Land dependence, which had been easing, could increase, raising the risk of agricultural expansion into forests, deforestation and reversal of previous forest gains. However, there could also be positive impacts – reduced wood demand could lessen pressure on forests, while conversion of forest for large-scale cultivation of commercial crops such as oil-palm, soybeans and rubber could slow as their prices fall.

It is impossible to know when the global economy will begin to recover. However, such crises also offer opportunities to chart new paths of development. The forest sector could benefit from the pursuit of a "green path" to development – through building up of natural resource capital (e.g. through afforestation and reforestation and increased investments in sustainable forest management), generation of rural employment and active promotion of wood in green building practices and renewable energy. Certainly, this change of path will require fundamental institutional changes, but the crisis may bring about greater willingness to accept and implement long-overdue reforms.

PART 1

Regional outlook

Changes in society, which have accelerated in recent decades with the rapid growth in information and communication technologies and globalization, are bringing about important changes in the forest sector at all levels. Demographic, economic, institutional and technological changes have altered the pattern of demand for forest products and services.

Considering the long-term nature of forestry, a better understanding of the potential directions of change is crucial to developing appropriate priorities and strategies for the future of the sector. In particular, the growing interaction of societies through globalization compels forestry professionals to acquire a broader perspective beyond national borders.

Part 1 has been developed largely based on ongoing and completed regional forest sector outlook studies. For each region, the drivers of change external to the forest sector are first examined in the areas of demographics, economy, policies and institutions, and science and technology. All demographic data are from World Urbanization Prospects: The 2007 Revision Population Database (UN, 2008a). All gross domestic product (GDP) and value-added figures are in 2006 constant United States dollars.

For each region, there is a section outlining an overall scenario of societal changes that can be expected up to 2030. Taking the predicted changes and previous trends into account, the future for the forest sector is then projected, tracing developments in:

- forest area, based – unless otherwise noted – on statistics from the Global Forest Resources Assessment 2005 (FAO, 2006a);

- forest management, addressing both natural and planted forests;
- wood products (production, consumption and trade), with projections based on econometric modelling, taking into particular account the changes in population, income and other parameters that influence demand;
- woodfuel, looking at both traditional woodfuel (fuelwood and charcoal) and modern biofuels, including the "wild card" of eventual commercial production of cellulosic biofuel;
- non-wood forest products (NWFPs), identifying broad patterns only, as the diversity of NWFPs in all regions makes it extremely difficult to provide a comprehensive outlook;
- environmental services of forests, including biodiversity conservation, climate change mitigation, protection against desertification and land degradation, watershed services and nature tourism. Protected areas are given as a rough indicator of conservation status; as recent reliable statistics do not exist by country for forest protected areas, figures are given for terrestrial protected areas (which include but are not limited to forests).

The aim has been to capture the main trends despite diversity within and among regions and countries which made the task a great challenge. An effort has also been made to provide a balanced perspective between short-lived changes, which usually capture most of the attention, and the less perceptible long-term changes, which are often neglected.

Africa

The African continent (Figure 1), consisting of 58 countries and areas (see Annex), contains highly diverse ecosystems. The continent accounts for 14 percent of the global population. The region's 635 million hectares of forests account for 21.4 percent of its land area. The Congo Basin hosts the second-largest contiguous block of tropical forest (Figure 2).

DRIVERS OF CHANGE

Demographics

Africa's population grew from 472 million in 1980 to 943 million in 2006 and is expected to rise to 1.2 billion by 2020 (Figure 3). Although the annual growth rate is declining (from 2.5 percent between 1990 and 2000 to a projected 2.1 percent between 2010 and 2020), the increase in absolute numbers implies further pressure on its resources.

Africa is urbanizing rapidly. By 2020, about 48 percent of the total population will be urban. However, with the exception of most of Northern Africa, Africa will remain largely rural in the next decade and beyond. The rural population is projected to increase by 94 million between 2005 and 2020.

HIV/AIDS will continue to affect the human and financial resources of a number of countries (Box 1).

Economy

In 2006, Africa accounted for about 2.3 percent of global GDP. Since 2000, the overall economic situation has improved. GDP growth rates have risen from 2.3 percent on average between 1990 and 1999 to more than 5 percent since 2000, reaching 6.2 percent in 2007 (IMF, 2008). High growth rates are likely to persist in the medium term (Figure 4). However, despite increased GDP growth, the per capita income growth rate remains low because of population growth (except in South Africa).

Low domestic savings and investments, uneven growth and skewed distribution of income remain concerns. Recent growth spurts are partly a consequence of the high prices of oil and other primary commodities.

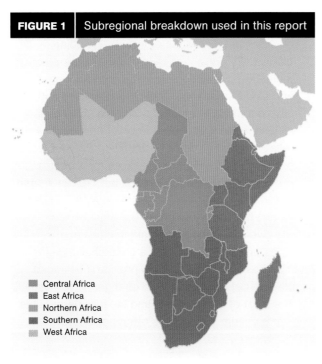

| FIGURE 1 | Subregional breakdown used in this report |

- Central Africa
- East Africa
- Northern Africa
- Southern Africa
- West Africa

NOTE: See Annex Table 1 for list of countries and areas by subregion.

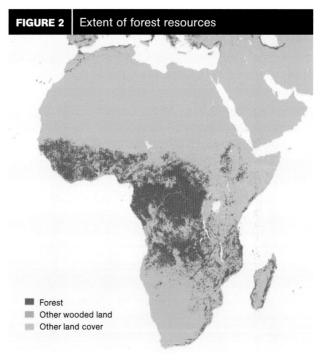

| FIGURE 2 | Extent of forest resources |

- Forest
- Other wooded land
- Other land cover

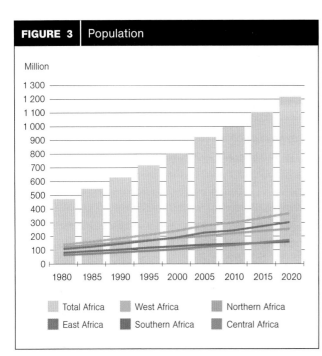

FIGURE 3 | Population

Million

Total Africa · West Africa · Northern Africa · East Africa · Southern Africa · Central Africa

SOURCE: UN, 2008a.

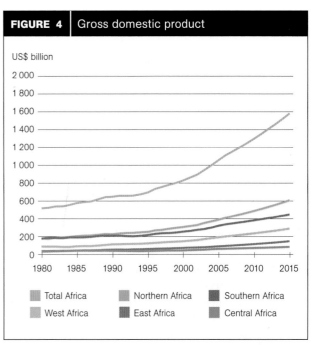

FIGURE 4 | Gross domestic product

US$ billion

Total Africa · Northern Africa · Southern Africa · West Africa · East Africa · Central Africa

SOURCES: Based on UN, 2008b; World Bank, 2007a.

BOX 1 | Impacts of HIV/AIDS

- Drastic decline in resources – human and financial – leaving less for long-term investments
- Increased dependence on forest products, especially those that are easy to collect
- Loss of traditional knowledge
- Shortage of skilled and unskilled labour – undermining forestry by affecting all key sectors such as wood industries, research, education, training, extension and forest administration
- Increased costs to industry on account of absenteeism and higher bills for treatment
- Reduced public-sector investment in forestry, as most governments will have to devote more of their budgets to health care and combating HIV/AIDS

SOURCE: FAO, 2003a.

Agriculture's share in gross value added has declined, from about 20 percent in the 1990s to 15 percent in 2006. However, agriculture is vital for livelihoods; it accounted for 70 percent of rural employment in 2005. Per capita productivity of agriculture is extremely low in comparison with other regions, and declining agricultural income has enhanced dependence on off-farm employment, including collection of fuelwood and NWFPs and production of charcoal.

Much of Africa's economic growth since 2000 has been driven by exports of primary commodities to the emerging Asian economies, and this is likely to continue. Africa's industries face major challenges, especially from increasing competition in domestic and global markets. Participation in global markets is expected to remain uneven because of limitations in policy and institutional frameworks, infrastructure, human resource development, the investment climate and competitiveness. African markets remain small and fragmented, although mechanisms for regional and subregional integration such as the

TABLE 1
Forest area: extent and change

Subregion	Area (1 000 ha)			Annual change (1 000 ha)		Annual change rate (%)	
	1990	2000	2005	1990–2000	2000–2005	1990–2000	2000–2005
Central Africa	248 538	239 433	236 070	−910	−673	−0.37	−0.28
East Africa	88 974	80 965	77 109	−801	−771	−0.94	−0.97
Northern Africa	84 790	79 526	76 805	−526	−544	−0.64	−0.69
Southern Africa	188 402	176 884	171 116	−1 152	−1 154	−0.63	−0.66
West Africa	88 656	78 805	74 312	−985	−899	−1.17	−1.17
Total Africa	**699 361**	**655 613**	**635 412**	**−4 375**	**−4 040**	**−0.64**	**−0.62**
World	**4 077 291**	**3 988 610**	**3 952 025**	**−8 868**	**−7 317**	**−0.22**	**−0.18**

NOTE: Data presented are subject to rounding.
SOURCE: FAO, 2006a.

Economic Community of West African States (ECOWAS) and the Southern African Development Community (SADC) are beginning to bear fruit.

Policies and institutions

Civil society's demand for transparency and good governance is bringing about fundamental changes in Africa. Decentralization of authority and participatory approaches to resource management are finding wider acceptance. However, conflicts undermine social and economic development in a number of countries.

Community involvement in natural resource management has a long history in Africa, and policy and legal changes in recent years have helped to accelerate devolution. However, forestry faces some enduring institutional difficulties such as:

- poor intersectoral linkages, with high-priority sectors such as agriculture, mining, industrial development and energy effectively having a greater impact on forests than forest policy;
- inconsistencies in laws governing the environment and those governing investments;
- poor governance and corruption in some countries;
- land tenure uncertainties, weak legal frameworks and other hindrances to the development of a competitive private sector;
- declining capacity of public forestry agencies, including research, education, training and extension.

Science and technology

With the exception of South Africa and some countries in Northern Africa, science and technology development in the region has been relatively slow, largely because of:

- low investments in science education and in research;
- the high share of economic activities remaining in the informal domain, which curbs interest to invest in innovations;
- a failure to develop and use Africa's strong base of traditional knowledge to deal with modern problems.

Furthermore, research and systematic enquiry do not tend to be fully mainstreamed in development planning and policy-making.

However, mobile communications and the Internet are improving access to information.

The forest sector reflects the general situation. Substantial efforts are required to revamp the institutional framework to strengthen the science and technology base of forestry. Otherwise, major breakthroughs are likely to bypass the African forest sector or at best will benefit only a small segment of the population.

OVERALL SCENARIO

Political and institutional developments will have the greatest influence on the forestry outlook and are the most uncertain (FAO, 2003a). A major transition – one that would favour balanced and equitable natural resource management – would depend on: improved efficiency and accountability in the public sector; greater inclusiveness, competitiveness and transparency in market institutions; and an informal sector (i.e. arrangements outside the public and market domains) that provides livelihood opportunities for the poor, especially where these are lacking in the formal sector. While improvements are being made in this direction, substantial efforts would be needed to effect a real turnaround before 2020. In most cases, a continuation along the current development path – a "business as usual" scenario – appears more likely.

OUTLOOK
Forest area

Although Africa holds only 16 percent of the global forest area, from 2000 to 2005 it lost about 4 million hectares of forests annually, close to one-third of the area deforested globally (Table 1). Most forest loss is taking place in countries with a relatively large forest area. To date, conversion to small-scale permanent agriculture has

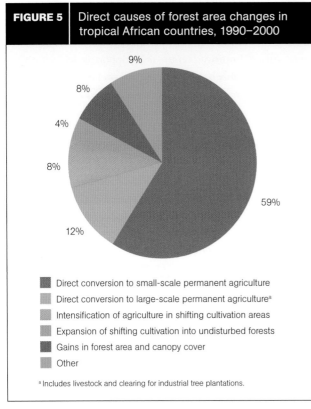

| FIGURE 5 | Direct causes of forest area changes in tropical African countries, 1990–2000 |

- Direct conversion to small-scale permanent agriculture
- Direct conversion to large-scale permanent agriculture[a]
- Intensification of agriculture in shifting cultivation areas
- Expansion of shifting cultivation into undisturbed forests
- Gains in forest area and canopy cover
- Other

[a] Includes livestock and clearing for industrial tree plantations.

SOURCE: FAO, 2001.

been the main contributor to forest loss (Figure 5), but investment in large-scale agriculture could become a major driver of deforestation in the future.

Forest loss is likely to continue at current rates. The growing demand for, and rising price of, food and energy will exacerbate the situation, especially as increased investments in infrastructure open up new areas. Climate change will also have an impact. Increasing frequency of droughts, declining water supplies and floods strain coping mechanisms at the local and national levels and undermine efforts to manage forests sustainably.

By subregion, the following picture is likely:
- An improvement in the economic situation in Northern Africa could help to reduce the pressure on land and reverse past trends of forest clearance, particularly in the Sudan. However, external investments in large-scale agriculture in response to high food prices could have a negative impact on forests.
- In East and Southern Africa, high population densities and high land dependence coupled with land-use conflicts and limited opportunities for economic diversification are likely to reduce forest area further.
- In Central Africa, low population densities, large expanses of land and improved accessibility may favour forest clearance for commercial and subsistence agriculture. Improved marketability of less-commercial species may lead to intensive

unsustainable logging, especially in the context of weak policies and institutions.
- In West Africa, rapidly growing urban demand for woodfuel and increasing agricultural demand is likely to result in continued reduction in forest cover.

Forest management

Natural forests continue to be the main source of wood supplies. The International Tropical Timber Organization (ITTO, 2006) found that only about 6 percent of the natural tropical production forests in the permanent forest estate of its ten African member countries were sustainably managed. Reduced-impact logging and harvesting codes are yet to find wide application, and investment in regeneration of logged areas is minimal.

Global concern about sourcing wood from sustainably managed areas is encouraging the adoption of certification in Africa. However, the extent of certification remains low because of the high transaction costs (Box 2).

Given the likelihood of a "business as usual" scenario, progress in implementing sustainable forest management is expected to be slow, primarily because of:
- the generally unfavourable investment climate;
- severe institutional, financial and technical constraints hindering forestry administrations' ability to manage logging concessions, which have often expanded so fast that governments cannot enforce rules and regulations and fully recover potential income;
- illegal activities and corruption;
- policies and institutional, technical and economic hurdles limiting wider adoption of community-based forest management, and a tendency to transfer only degraded forests to local communities, which lack the investment capacity to rehabilitate them.

All of the above favour unsustainable exploitation. Depending on how community capacity is built up, some progress in sustainable forest management is expected in the savannah woodlands, especially in East and Southern Africa, although it may be hindered by low returns from these forests.

With an estimated 14.8 million hectares of planted forests (FAO, 2006b), Africa accounts for only about 5 percent of the

| BOX 2 | Forest certification in Africa |

Of the 306 million hectares of certified forests in the world (June 2007), Africa accounts for about 3 million hectares (about 1 percent). Most of Africa's certified forests are planted forests, and about half are in South Africa.

SOURCE: ITTO, 2008.

global total. Of this, about 3 million hectares were planted for protection and the rest for production of wood and non-wood forest products (e.g. gum arabic). Most of Africa's wood is produced from natural forests; investments in planted forests have occurred mainly in countries with relatively low forest cover (Algeria, Morocco, Nigeria, South Africa and the Sudan). Average annual planting in Africa from 1990 to 2005 was estimated at about 70 000 ha, less than 2 percent of the global planting rate. In several countries, planted forest area has declined in recent years.

With the exception of South Africa, most planted forests are established and managed by public forestry agencies. Expansion of forest planting and intensive management for production will largely depend on plantation profitability as perceived by the private sector, taking into account the global demand for wood products. Realization of the potential in some of the countries requires significant improvements in the policy and institutional framework, including landownership.

Growing demand for wood has encouraged farm planting in most countries, and trees outside forests have become an increasingly important source of timber and fuelwood (Box 3). This trend is expected to intensify in the coming years. The potential of farm planting to supply

industrial roundwood and the constraints in obtaining land for large-scale planted forests have encouraged industries to enter into partnership with communities, for example in South Africa. Improved tenure and supportive legislation could considerably boost tree planting on farms, as is already happening in many countries (e.g. Ghana, Kenya and Uganda).

Wood products: production, consumption and trade

Africa produced 19 percent of global roundwood in 2006. Roundwood production increased slightly between 1995 and 2006, from 568 to 658 million cubic metres, roughly corresponding to the proportion of area under forests. However, woodfuel accounts for about 90 percent of roundwood production. The higher is the degree of processing, the lower is the share of Africa's contribution. Thus, while Africa accounts for more than one-quarter of global woodfuel production, its share in other wood products is very low (Table 2).

South Africa produced about 20 percent of Africa's industrial roundwood in 2006, largely from planted forests. Nigeria produced another 13 percent.

In view of the limited extent of forests and their low productivity, Northern Africa produces less than 6 percent of Africa's industrial roundwood and, hence, is highly dependent on imports.

In recent years, production of industrial roundwood from natural forests has declined in most West African countries and increased in Central African countries (Cameroon, the Democratic Republic of the Congo and Gabon) as large concessions have been awarded.

Some countries have imposed restrictions on the export of logs in order to encourage domestic processing, but this has not necessarily had the intended result of value addition. At best, it has led to some investments in preliminary processing.

Gross value added increased from about US$12 billion in 2000 to US$14 billion in 2006 (Figure 6). Increases have been entirely in roundwood production; value addition in wood processing and pulp and paper has stagnated.

BOX 3	Trees outside forests

Trees grown on homestead farms, in woodlots and on communal lands are an important source of wood and other products. In humid-zone West African countries, such as Burundi, Rwanda and Uganda in particular, trees grown in home gardens meet most household needs for fuelwood and timber. In many cash-crop systems, trees are grown for shade and eventually provide wood – an example is *Grevillea robusta* in tea plantations in Kenya. In the Sudan, *Acacia senegal*, the source of gum arabic, is largely grown in agroforestry systems, although some mechanized farms have also undertaken its cultivation on a larger scale in recent years.

SOURCE: FAO, 2003a.

TABLE 2
Wood product output, 2006

Product	Global	Africa	Share (%)
Industrial roundwood *(million m³)*	1 635	69.0	4
Sawnwood *(million m³)*	424	8.3	2
Wood-based panels *(million m³)*	262	2.5	1
Pulp for paper *(million tonnes)*	195	3.9	2
Paper and paperboard *(million tonnes)*	364	2.9	1
Woodfuel *(million m³)*	1 871	589.0	46

SOURCE: FAO, 2008a.

Industrial roundwood production is expected to grow in the next two decades (Table 3), and some of the subregional shifts will become more pronounced. Southern Africa's share of industrial roundwood production (which is primarily attributed to South Africa) is expected to rise, considering potential increases in logging (especially in Angola and Mozambique). Marginal increases are expected in West Africa and Northern Africa; a decline is expected in East Africa. Central Africa is emerging as a major producer of industrial roundwood. Realizing the potential demand will depend on increases in income and overall social and economic development.

Africa's share in the global wood products trade is extremely low (Table 4) and is geared to the production of low-value-added items (with the exception of South Africa). Intraregional trade in wood products is also low. Between 1980 and 2006, Africa's total wood products exports increased from US$1.6 billion to US$4 billion, while its share of the global total (now in excess of US$200 billion) declined. Realizing Africa's potential in the wood products industry depends on the creation of a favourable policy and institutional environment and on improving competitiveness.

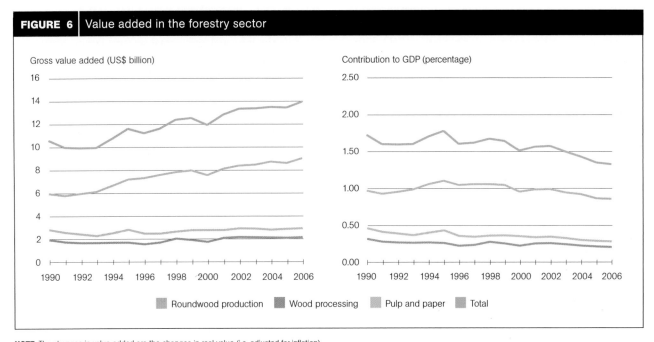

FIGURE 6 | Value added in the forestry sector

Gross value added (US$ billion)

Contribution to GDP (percentage)

Roundwood production   Wood processing   Pulp and paper   Total

NOTE: The changes in value added are the changes in real value (i.e. adjusted for inflation).
SOURCE: FAO, 2008b.

TABLE 3
Production and consumption of wood products

Year	Industrial roundwood (million m³)		Sawnwood (million m³)		Wood-based panels (million m³)		Paper and paperboard (million tonnes)	
	Production	**Consumption**	**Production**	**Consumption**	**Production**	**Consumption**	**Production**	**Consumption**
2000	69	64	8	11	2	2	4	5
2005	72	68	9	12	3	3	5	7
2010	81	77	10	15	3	4	7	10
2020	93	88	11	19	4	4	9	14

SOURCE: FAO, 2008c.

TABLE 4
Africa's share of trade in wood products, 2006

Product	Imports as % of global import value	Exports as % of global export value	Imports as % of quantity consumed in Africa	Exports as % of quantity produced in Africa
Industrial roundwood	0.7	8.4	1.0	6.0
Sawnwood	3.3	3.0	45.0	23.0
Wood-based panels	1.4	1.9	45.0	37.0
Pulp for paper	0.8	1.0	26.0	36.0
Paper and paperboard	2.5	0.6	51.0	12.0

SOURCE: FAO, 2008a.

Woodfuel

Traditional energy sources (mainly biomass) dominate the energy sector, especially in sub-Saharan Africa, where only 7.5 percent of the rural population has access to electricity (World Energy Council, 2005). As household incomes and investment in appropriate alternatives remain low, wood is likely to remain an important energy source in Africa in the coming decades (FAO, 2008d). Forecasts made in 2001 suggested a 34-percent increase in woodfuel consumption from 2000 to 2020 (Figure 7). However, the rise in fuel prices in the past two years suggests that this increase is likely to be even greater. The share of woodfuel in the total energy supply is likely to decline, but the absolute number of people dependent on wood energy is predicted to grow (FAO, 2008d).

Although woodfuel supply and demand are balanced at the aggregate level, there are areas of acute deficit, resulting in unsustainable removals, particularly around urban centres. Most countries have attempted to boost supply through improved management of forests and woodlands and the establishment of woodfuel plantations, and to reduce demand by promoting more-efficient cooking devices and alternative fuels.

Global interest in biofuels as a result of rising fossil fuel prices has increased investments in biofuel development, for example through the planting of *Jatropha* species. It is uncertain whether these investments will provide a long-term solution to Africa's energy problems, and there are growing concerns about adverse implications for food security.

Non-wood forest products

African NWFPs (gums and resins, honey and beeswax, dying and tanning materials, bamboo and rattan, bushmeat, fodder and a considerable number of medicinal plants) are largely used for subsistence and traded informally. Thus, their livelihood contribution and local significance exceed that which may be apparent from official statistics (Shackleton, Shanley and Ndoye, 2007).

With increased opportunities for local, regional and international trade, the NWFP sector in Africa is undergoing perceptible changes. African governments are increasingly developing policies and legislation aimed at formalizing NWFP value chains. Of particular significance is the emergence of markets for "ethnic foods", medicinal plants and natural or organic goods, such as honey, beeswax and shea butter (Box 4). Several products that are traded nationally and internationally straddle the informal and formal sectors. For example, collection from the wild may remain in the informal sector, while processing and trade are in the formal sector.

In view of the wide range of products and end uses, it is difficult to make a widely applicable forecast, but the outlook is likely to include:

- subsistence consumption of most products with little attention to management of the resources;
- overexploitation and depletion of some wild resources collected for commercial products;
- further pressure on bushmeat resources as a result of increased population;
- domestication and commercial cultivation and processing of a small number of products by entrepreneurs or local communities;

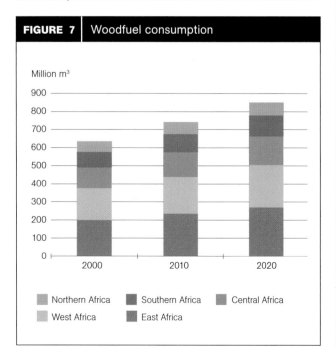

FIGURE 7 | Woodfuel consumption

Million m³

Legend: Northern Africa, Southern Africa, Central Africa, West Africa, East Africa

---

BOX 4	Shea-butter-based cosmetic products

Cosmetics such as oils, creams and dyes represent one of the fastest-emerging global markets for non-wood forest products. Shea butter, derived from the fruit of the shea tree (*Butyrospermum parkii* or *Vitellaria paradoxa*) and commonly known as karité, is one of the most popular ingredients in skin care today. The shea tree is found only in Africa's Sahel belt, and it is estimated that 3 million rural African women are involved in the export of shea products, which were valued at US$100 million for 2007–2008. In Burkina Faso, karité is the second-largest export item after cotton, and several projects are focusing on developing the sector. For example, Burkina Faso's Project Karité organizes local women's associations that collect and process shea nuts and karité for international markets. As the women run their own operations, activities typically comply with "fair trade" conditions. In addition, most of the small village enterprises supported by TREE AID in Burkina Faso are based on karité.

- growing demand from niche markets for certified and fair-trade products (Welford and Le Breton, 2008).

Environmental services of forests

Under a "business as usual" scenario, forest biodiversity loss is likely to continue. Efforts to reverse the situation should build on the successes of community management initiatives such as the Communal Areas Management Programme for Indigenous Resources (CAMPFIRE) in Zimbabwe (Frost and Bond, 2008). Biodiversity conservation also needs to be addressed outside protected areas and integrated into key economic activities.

Protected areas currently cover about 320 million hectares (11 percent of the region's land area), but Africa's investment and staffing in park management remain the lowest in the world. Major challenges to protected area management include increasing human–wildlife conflict (FAO, 2008e) and resource-use conflicts, which often worsen in the event of drought. In Kenya, the United Republic of Tanzania and Zimbabwe, among others, local communities are involved in managing protected areas or tourism facilities for a share of the income. Leasing of protected areas for management has not yet taken hold in Africa.

Climate change will have significant impacts on African economies and on the forest sector. The Clean Development Mechanism (CDM) of the Kyoto Protocol and recent initiatives for reducing emissions from deforestation and forest degradation (REDD) open up new funding opportunities. Hitherto, Africa has not benefited much from the CDM or voluntary carbon markets (Box 5), suggesting that vigorous efforts are needed to address the inadequate technical capacity and policy and institutional constraints if the region is to be able to take advantage of REDD.

Acute water scarcity affects both rural and urban areas in several African countries, and it is expected to worsen as demand increases. Poor watershed management has resulted in heavy siltation and diminished storage capacity in many reservoirs. Fragmented responsibilities and conflicting uses are the main constraints in watershed management, especially for the several transboundary watersheds in the region. The main challenges are to adopt integrated land use and to develop institutional arrangements linking upstream land users and downstream water users.

Interest in a market approach for provision of watershed services is just beginning to grow. The region has only two programmes of payment for environmental services (PES) involving watersheds, both in South Africa, and neither is strictly market-based as they depend on general tax revenue. Several other initiatives are in the

BOX 5	Carbon markets in Africa: an overview

- Global total registered Clean Development Mechanism (CDM) projects to 30 April 2008: 1 068
- CDM projects in Africa: 25 (2.3 percent of the total), most in South Africa (where institutional capacity is relatively well developed)
- Approved afforestation/reforestation projects in Africa: none (in the world: one [in China])
- Africa's share in voluntary carbon markets: 2 percent of the volume transacted in 2007, with the highest-priced credits because of high transaction costs
- Africa's share in voluntary carbon markets in land use, land-use change and forestry in 2007: 5 percent of the global total

SOURCE: Hamilton et al., 2008.

planning stage. The main challenges for such schemes are users' inability to pay for watershed services, high transaction costs and institutional deficiencies (Dillaha et al., 2007).

Desertification and land degradation affect most African countries and are expected to worsen with climate change, grazing expansion and increasing pressure to cultivate marginal lands. Trees planted in windbreaks and shelterbelts protect agricultural land and infrastructure. Addressing desertification and land degradation requires an integrated approach to agriculture, animal husbandry and forestry – as adopted in regional and subregional projects such as the Green Wall for the Sahara Initiative (see UNU, 2007) and the TerrAfrica partnership (TerrAfrica, 2006). Almost all countries in the region are signatories to the United Nations Convention to Combat Desertification (UNCCD) and have developed national action plans (often with external support). However, economic and institutional constraints limit the ability of governments, the private sector and communities to address the challenges systematically.

Nature-based tourism and emerging private-sector-led and community-based ecotourism initiatives, primarily centred on protected areas, make a significant contribution to African economies. The rich wildlife is a major source of income and employment. Africa has considerable potential to take advantage of growth in global tourism. However, the overall trend of continued deforestation and forest degradation implies a diminishing supply of forest-derived environmental services. Whether increasing awareness of the environmental services provided by African forests will influence their conservation depends on the costs involved.

SUMMARY

The forest situation in Africa presents enormous challenges, reflecting the larger constraints of low income, weak policies and inadequately developed institutions. Success stories exist but remain isolated because of fundamental economic and institutional weaknesses. Obstacles include:

- high dependence on land and natural resources and scant investment in development of human resources, skills and infrastructure;
- the low level of value addition in the economy, including the forest sector;
- the vastness of the informal sector, stemming from the weaknesses in the public sector and market mechanisms.

Focusing on the unique products and services required locally and globally and strengthening local institutions can be important ways of addressing forest resource depletion. Such efforts should build on successful experience with locally based sustainable resource management integrating agriculture, animal husbandry and forestry, and take advantage of local knowledge. The growing demand for environmental services – especially biodiversity and carbon sequestration – provides a particular opportunity for Africa.

Asia and the Pacific

The Asia and the Pacific region (Figure 8), consisting of 47 countries and areas, is home to more than half of the world's population and has some of the most densely populated countries in the world. It has 18.6 percent of the world's forest area in a wide array of ecosystems including tropical and temperate forests, coastal mangroves, mountains and deserts (Figure 9). Rapid socio-economic changes in the region are having profound impacts on all sectors, including forestry. While wood products demand is increasing, so is the demand for environmental services of forests.

DRIVERS OF CHANGE
Demographics

The population of Asia and the Pacific is projected to reach 4.2 billion by 2020, an increase of 600 million from 2006 (Figure 10). The annual population growth rate in Japan is close to zero and declining, but in several countries – particularly low-income countries – the growth rate exceeds 2 percent.

Population density in the region varies enormously, from fewer than 2 people per square kilometre in Mongolia to more than 1 000 people per square kilometre in Bangladesh and to more than 6 300 people per square kilometre in completely urban Singapore.

The urban population in Asia and the Pacific is expected to rise from 38 percent in 2005 to 47 percent in 2020. It is increasing especially rapidly in China – by 2020, China's urban population is expected to have grown by 230 million and its rural population to have decreased by 122 million relative to 2005 figures. South Asia, where 65 percent of the population is rural, is expected to remain the least urbanized subregion.

Another important demographic change is the ageing of the population. In Australia, Japan, Malaysia, New Zealand and Thailand, more than 15 percent of the population is over 65 years old; in Japan, more than one-quarter of the population is over 60 years old. The reduction in the proportion of working-age adults in these countries, as well as in China (where a strict population policy is implemented), will have important implications for productivity and the demand for goods and services.

FIGURE 8	Subregional breakdown used in this report

- East Asia
- South Asia
- Southeast Asia
- Oceania

FIGURE 9	Extent of forest resources

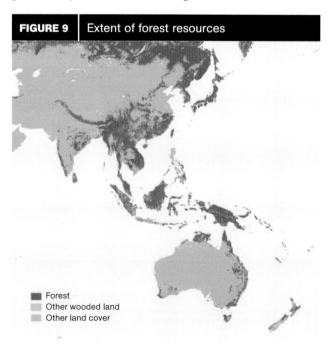

- Forest
- Other wooded land
- Other land cover

NOTE: See Annex Table 1 for list of countries and areas by subregion.

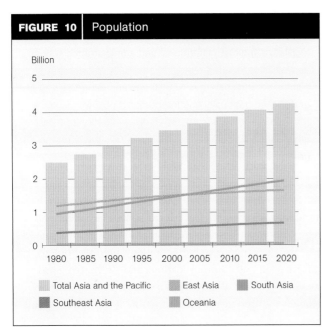

FIGURE 10 | Population

Billion

Total Asia and the Pacific   East Asia   South Asia
Southeast Asia   Oceania

SOURCE: UN, 2008a.

Economy

Asia and the Pacific has the fastest economic growth of all regions. China and India, which account for two-thirds of the region's population, have registered annual GDP growth rates of 8–11 percent during the past decade. While some slowdown is possible, most countries are expected to have growth rates well above the global average (Figure 11).

However, despite notable poverty reduction since the 1990s, the region still has 640 million people living on less than US$1 per day (UNESCAP, 2007). With poverty more pervasive in forested areas, many people depend in large part on forests for their livelihood.

In most developing countries in the region, the manufacturing and services sectors are growing rapidly, with a corresponding decline in the share of agriculture in income and employment (UN, 2006a; FAO, 2007b). These changes in the structure of the economy will have different effects on forests and forestry depending on their pace:

- Several countries in the region will remain largely dependent on agriculture. High population growth and continued dependence on land will raise the

pressure on forests, especially in densely populated countries. Efforts to improve agriculture in response to recently escalating food prices could increase the impact on forests.
- In countries where industrialization is reducing the pace of agricultural expansion, other factors such as mining, infrastructure development and urbanization, as well as plantation crops, are becoming important causes of forest clearance.
- Some countries have become, or are becoming, knowledge economies, largely focused on technology and services. With high incomes, most of the primary commodities are imported and dependence on forests is reduced. Forests are then used primarily for the provision of environmental services.

Globalization has played an important role in the region's rapid economic growth and will become more pronounced in the coming years, with continued impact on the forest sector, including increased transnational investments. Relative political stability, large markets, high investments in human resources, regional and subregional trade and economic cooperation agreements, improved

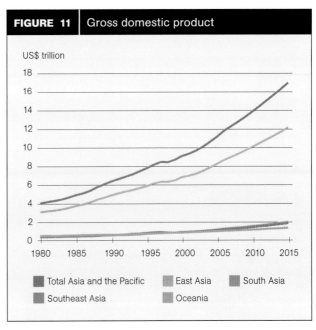

FIGURE 11 | Gross domestic product

US$ trillion

Total Asia and the Pacific   East Asia   South Asia
Southeast Asia   Oceania

SOURCES: Based on UN, 2008b; World Bank, 2007a.

transportation infrastructure and rapid development of information and communication technologies have all promoted globalization.

Policies and institutions

Important changes under way in the policy and institutional arena in Asia and the Pacific include:

- changes in policies and legislation enabling greater involvement of diverse stakeholders in forestry, especially through privatization and community participation, including the restoration of rights to indigenous communities (Box 6);
- improvement in tenure conditions providing more incentive for landowners to grow trees;
- increased corporate investments in forestry, often through partnerships;
- greater involvement of civil-society organizations in policy formulation, forest management, research, extension and awareness generation;
- weakening of the authority of public forestry agencies, largely because of the emergence of other players.

Governance problems encumber the forest sector in some countries, often those with the most forests. Combating corruption and illegal logging has been a focus of recent national and international efforts (Box 7). Conflicts disrupt forest management in several countries, and these could escalate as pressures on natural resources increase, especially if effective institutional arrangements are not in place to resolve them.

Science and technology

The region has been at the forefront of the development and adoption of green revolution technologies, which have slowed or even reversed the horizontal expansion of

---

BOX 6	Restoration of rights to indigenous communities

Of the estimated 210 million to 260 million indigenous people in Asia and the Pacific, about 60 million are forest-dependent. Many countries have policies and laws to remedy their marginalization (e.g. Australia, India, Malaysia, New Zealand, Papua New Guinea and the Philippines). For example, the Scheduled Tribes and Other Traditional Forest Dwellers (Recognition of Forest Rights) Act enacted by India in 2006 recognizes the rights of traditional forest-dwelling communities, including title over land that they have been cultivating (up to a maximum of 4 ha per family) and the right to collect and use non-wood forest products.

SOURCE: Asia Forest Network, 2008.

BOX 7	Forest law enforcement and governance in Asia

In Asia, multilateral arrangements on forest law enforcement and governance (FLEG) target explicit improvements in reducing corruption and illegal activities in, and associated with, forests and forestry. The East Asian FLEG process emerged from a series of multistakeholder consultations in 2001. A ministerial FLEG meeting held in Bali, Indonesia, in 2001 affirmed commitments to eliminate illegal logging and associated illegal trade and corruption. It also developed a comprehensive list of actions – encompassing political, legislative, judicial, institutional and administrative actions as well as associated research, advocacy, information disclosure and sharing of knowledge and expertise – to be undertaken nationally and internationally. However, while the FLEG process has helped to draw attention to forest governance, it is difficult to ascertain its impacts on the ground.

---

agriculture. Technological advances have enhanced the region's competitiveness in the manufacturing and services sectors. Investments in biotechnology, nanotechnology, information and communications and alternative energy technologies will all have important impacts on forestry. However, differences in the adoption of technologies will persist among countries, sectors and subsectors.

OVERALL SCENARIO

The Asia and the Pacific region is extremely diverse. Countries, or even areas within countries, are likely to follow one of three main development paths.

In the rapidly emerging industrial economies, continued industrialization will result in an expanded middle class. A consequent increase in the demand for food, fuel, fibre and environmental services will exert tremendous pressure on the natural-resource-rich countries in and outside the region. Agricultural expansion will slow; non-agricultural land uses such as mining and urban expansion may continue to place pressure on forests.

In the agrarian societies, agriculture will remain the mainstay of livelihoods and may even expand in the context of high population growth rates. Increasing global and regional demand for food, fuel and fibre, especially from rapidly industrializing countries, could be either an opportunity or a challenge depending on the state of governance and institutional development.

In the high-income, postindustrial societies, growth will be based on technologically advanced manufacturing and the provision of high-quality services. Populations will be relatively stable (and in some cases declining) and technically skilled. Improving the quality of the environment will be a major concern, and high income will provide the necessary means to do so.

OUTLOOK

Forest area

Asia and the Pacific had 734 million hectares of forest in 2005, about 3 million hectares more than in 2000 (Table 5). However, this increase was largely a result of the high afforestation rate in China, masking significant loss of natural forests in a number of countries; in the region as a whole, 3.7 million hectares were lost annually between 2000 and 2005.

Considering the two dominant development paths – rapid economic growth through industrialization and agriculture remaining the mainstay of livelihoods – forest loss is likely to continue in most countries in the next two decades at more or less the current rates. Some countries have reversed their trends of forest loss, but the countries with the most deforestation are unlikely to be able to do so. Expansion of large-scale commercial crops will be the most important driver of deforestation in the region (Figure 12), especially as oil-palm cultivation expands to meet the growing demand for biodiesel and foodgrain prices rise. In addition, in the more populous countries, especially those in South Asia, forest degradation will be a major problem, stemming from unsustainable collection of wood and non-wood forest products and from grazing.

Forest management

In natural forests managed for wood production, the region has made major efforts to implement sustainable forest management through such measures as reduced-impact logging and the use of certification to target niche markets, with many success stories (see FAO, 2005a). ITTO (2006) reported 14.4 million hectares of natural tropical production forests in the permanent forest estate of its ten member countries in the region as sustainably managed, mostly in India, Indonesia and Malaysia.

With increasing wood production from planted forests, the area of natural forests managed for wood supply has declined, partly because of the complexity and higher costs of natural forest management. Some countries have imposed outright logging bans, setting natural forests aside for their environmental values. However, where institutional arrangements are weak, unsustainable and often illegal logging is likely to continue, depressing the economic viability of sustainable forest management.

Asia and the Pacific has 136 million hectares of planted forests, nearly half of the global total (Table 6). However, their productivity is far short of their potential.

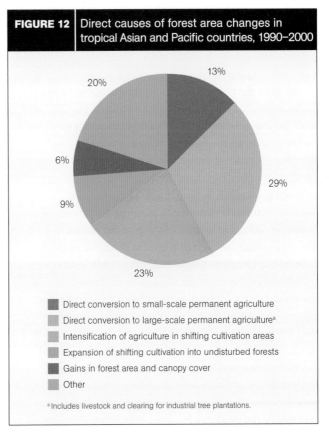

| FIGURE 12 | Direct causes of forest area changes in tropical Asian and Pacific countries, 1990–2000 |

- Direct conversion to small-scale permanent agriculture
- Direct conversion to large-scale permanent agriculture[a]
- Intensification of agriculture in shifting cultivation areas
- Expansion of shifting cultivation into undisturbed forests
- Gains in forest area and canopy cover
- Other

[a] Includes livestock and clearing for industrial tree plantations.

SOURCE: FAO, 2001.

TABLE 5

Forest area: extent and change

Subregion	Area (1 000 ha)			Annual change (1 000 ha)		Annual change rate (%)	
	1990	2000	2005	1990–2000	2000–2005	1990–2000	2000–2005
East Asia	208 155	225 663	244 862	1 751	3 840	0.81	1.65
Oceania	212 514	208 034	206 254	−448	−356	−0.21	−0.17
South Asia	77 551	79 678	79 239	213	−88	0.27	−0.11
Southeast Asia	245 605	217 702	203 887	−2 790	−2 763	−1.20	−1.30
Total Asia and the Pacific	**743 825**	**731 077**	**734 243**	**−1 275**	**633**	**−0.17**	**0.09**
World	**4 077 291**	**3 988 610**	**3 952 025**	**−8 868**	**−7 317**	**−0.22**	**−0.18**

NOTE: Data presented are subject to rounding.
SOURCE: FAO, 2006a.

TABLE 6
Planted forest area change

Year	Extent of planted forests			Global total	Annual change in Asia and the Pacific
	Productive	**Protective**	**Total**		
	(million ha)				
1990	67	36	103	209	–
2000	78	41	119	247	1.4
2005	90	46	136	271	2.8

SOURCE: FAO, 2006b.

TABLE 7
Production and consumption of wood products

Year	Industrial roundwood *(million m³)*		Sawnwood *(million m³)*		Wood-based panels *(million m³)*		Paper and paperboard *(million tonnes)*	
	Production	**Consumption**	**Production**	**Consumption**	**Production**	**Consumption**	**Production**	**Consumption**
2005	273	316	71	84	81	79	121	128
2020	439	498	83	97	160	161	227	234
2030	500	563	97	113	231	236	324	329

SOURCE: FAO, 2008c.

Most of the planted forests are in Australia, China, India, Indonesia, New Zealand, the Philippines, Thailand and Viet Nam. Investments in planted forests, especially by the private sector, have increased in the past two decades. As more of the natural forests are excluded from production, planted forests are becoming the mainstay of wood production in the region. There has also been substantial investment in planting for protective purposes; almost one-third of the planted forests in the region have been established for environmental protection, mostly in China and India (FAO, 2006b).

However, the scope for expansion of planted forests for production is limited, especially with current wood prices. Water availability is already a major constraint and will be more problematic in the future. The costs of productive land are steep, inflated by high agricultural prices and demand for biofuel feedstocks. Although marginal land is extensively available, it requires high investments. Thus, future wood supply will depend on improving the productivity of existing planted forests and on encouraging farm forestry as an important source of wood, including for large-scale industrial processing (Box 8).

Wood products: production, consumption and trade

Regionally, large increases in industrial roundwood consumption and production are projected to 2020 (Table 7). China, India and other emerging economies will account for much of the growth in consumption. Trends in industrial roundwood imports are in contrasting directions. Net imports to the advanced industrialized economies (especially Japan) have

---

BOX 8	Farm forestry

Trees are an integral part of homestead farming systems in many Asian countries, particularly Bangladesh, Indonesia, the Philippines, Sri Lanka and certain parts of India. Past investments in social or community forestry have helped to make farms important sources of wood supply. Several industries have established partnership arrangements with farmers to source wood supplies from farms. Farm forestry is expected to continue to expand as a result of:

• improving security of land tenure;
• declining profitability of agriculture, which encourages farmers to invest in forest crops (which are less labour-intensive than agriculture);
• increasing demand for wood products and consequent increases in their prices, making farm forestry more profitable.

---

declined, while those to the emerging economies (China and India) have greatly increased as a result of surging domestic demand and declines in domestic supply caused by logging bans.

Growth in demand for wood products (Figure 13) will largely be a continuation of recent trends and will be similar to the global outlook (see Part 2), with substantial expansion expected in the consumption of wood-based panels and paper and paperboard, and more modest growth in sawnwood consumption. Sawnwood and plywood will continue to account for most of the consumption of solid wood products, although some substitution of

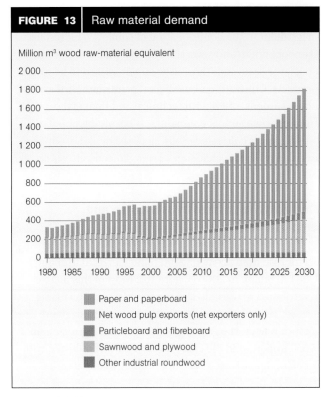

FIGURE 13 | Raw material demand

Million m³ wood raw-material equivalent

Paper and paperboard
Net wood pulp exports (net exporters only)
Particleboard and fibreboard
Sawnwood and plywood
Other industrial roundwood

SOURCE: FAO, 2008c.

reconstituted panels for sawnwood and plywood is expected. Consumption of paper and paperboard is expected to grow markedly, and recovered paper and wood produced in planted forests of fast-growing species will provide most of the fibre used in their production.

Rapid economic growth has boosted the region's share in the global trade in wood products, especially in the past two decades. Rising prosperity generally implies higher disposable income, increasing the demand for products and consequently imports (Figure 14). China accounts for a large part of the growth in trade; its total wood products imports rose from US$5.4 billion in 1990 to US$20.6 billion in 2006. India's wood products imports have also increased notably, from about US$587 million in 1990 to US$2.4 billion in 2006. China's recovered paper imports (mainly from the United States of America) grew from 5 million tonnes in 2000 to 16.7 million tonnes in 2006.

The region is also becoming an important exporter of wood products, with an increasing share of high-value products. Most remarkable is the emergence of China as the leading global exporter of furniture, overtaking some of the traditional furniture producers in Europe. Since 2005, Viet Nam has also emerged as a main exporter of wooden furniture.

The trends in demand and trade have several interesting implications for the future of forests in the region:
- The boom in demand creates opportunities for forested countries, but also challenges for sustainable forest management and control of illegal logging,

especially in countries with weak institutions and poor governance.
- Growth in trade may have impacts on forest management outside the region.
- Some of the demand may be met through improvements in efficiency.

Woodfuel

Almost three-quarters of the wood produced in Asia and the Pacific is burned as fuel. In South and Southeast Asia, woodfuel's share in total wood production is 93 and 72 percent, respectively. In contrast, woodfuel accounts for less than 1 percent of the wood produced in Japan.

Woodfuel consumption in the region declined between 1980 and 2006 from about 894 million to 794 million cubic metres. South Asia was the only subregion that registered an increase.

As incomes and urbanization increase, woodfuel will be substituted with electricity, kerosene and gas. This is already evident in most of Asia and the Pacific, although there are some differences in the predicted trends among subregions (Figure 15). For example, South Asian woodfuel consumption is expected to grow and then start to decline from around 2015. However, rising fossil fuel prices could lead to a different scenario, and the predicted fuel switching may not take place. In some cases, there could be a shift back to woodfuel, with consequences of increased collection and forest degradation.

Recent oil price increases have already led to substantial public and private investments in biofuel production. Oil-yielding species such as *Jatropha curcas* are being planted on degraded land for biodiesel production. As biodiesel is mainly used only in transportation, this

FIGURE 14 | Wood products imports

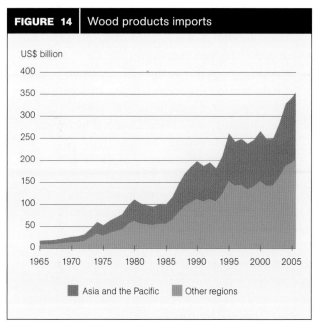

US$ billion

Asia and the Pacific   Other regions

SOURCE: FAO, 2008a.

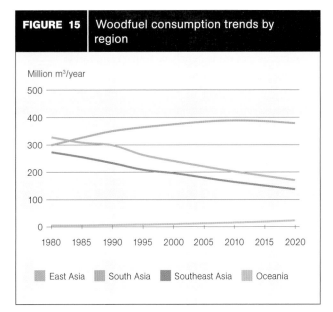

FIGURE 15 | Woodfuel consumption trends by region

Million m³/year

Legend: East Asia, South Asia, Southeast Asia, Oceania

SOURCE: FAO, 2003b.

development may not alleviate the traditional woodfuel problem.

If cellulosic biofuel production becomes commercially viable, the demand for wood as a source of energy will increase significantly.

Non-wood forest products

Non-wood forest products from the region are diverse – food, medicines, fibres, gums, resins, cosmetics and handicrafts. Most are used for subsistence, collected and consumed locally or traded in limited quantities. More than 150 NWFPs from Asia and the Pacific are traded internationally, although apart from bamboo and rattan the quantities are usually small. Increasing interest in "natural products", owing to their perceived health and environmental benefits, is drawing attention to the multitude of NWFPs commonly used by local communities.

The consumption of many subsistence NWFPs is likely to fall in the long term because of:

- declining supply from the wild, largely because of reduction in forest cover and poor management;
- development of synthetic materials and their substitution for NWFPs as a result of increasing incomes and consumer access;
- the decreasing attractiveness of NWFP collection relative to more remunerative and less arduous occupations available when incomes rise.

Several NWFPs – especially medicinal plants – have been commercialized and are traded nationally and globally. Increasing demand has led to their intensive collection and to depletion of wild stock. Products from open-access public forests are particularly vulnerable. In many cases, collection and trade are informal, offering minimal financial benefits to collectors.

Declining supply from the wild has led to substantial investment in the domestication of some NWFP resources. Bamboo, rattan and several medicinal plants are grown on a large scale and, thus, have largely ceased to be forest products. Cultivation of medicinal plants on farms and in home gardens, often with technical and financial support from pharmaceutical companies, is becoming popular. As with most cultivated crops, periodic demand–supply imbalances create challenges for organized cultivation of NWFPs.

Contribution of forestry to income and employment

In absolute terms, the value added generated by the forestry sector rose from about US$100 billion in 2000 to about US$120 billion in 2006 (Figure 16). Most of this increase is attributed to the pulp and paper and wood-processing sectors, while wood production has remained stagnant. This pattern reflects the growing dependence of the region on wood imports and the changing structure of industry, with greater emphasis on more value-adding manufacturing. However, the share of forestry in GDP and employment continues to decline (Figure 17), largely because of the much faster growth of other sectors of the economy.

Environmental services of forests

The current situation and outlook for the provision of environmental services from forests are extremely varied in the region. National policies and strategies are focusing increasingly on environmental services of forests, and several countries have imposed logging bans in response to catastrophic events such as flooding and landslides. The provision of environmental services relies more on regulatory than market approaches.

The region has a long history of protected area management, but the control of illegal encroachment is often a challenge. Shrinking habitats are increasing human–wildlife conflicts, and trafficking in animals and animal parts is soaring. Declines have been reported for flagship species such as tiger and rhinoceros. In view of the continued degradation of protected areas, increasing emphasis is being given to participatory management, enabling local communities to benefit from protected areas, for example through ecotourism.

The region has extensive and highly fragile dry lands. Increasing socio-economic pressures have led to cultivation of marginal lands and overgrazing, which in tandem with climate variations are accelerating desertification. Many countries (e.g. China, India, Mongolia and Pakistan) implement tree planting and integrated land-use systems to combat degradation and desertification, including windbreaks and shelterbelts to protect agricultural land.

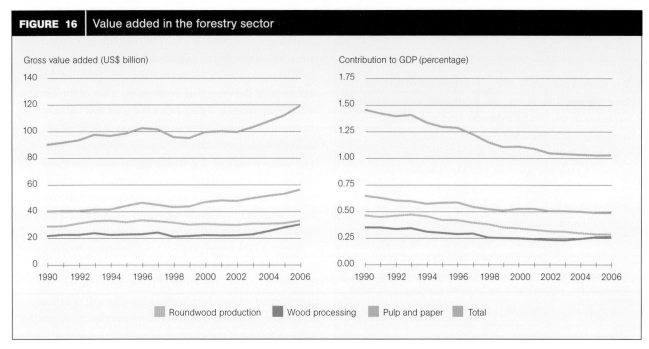

FIGURE 16 | Value added in the forestry sector

Gross value added (US$ billion)

Contribution to GDP (percentage)

Roundwood production   Wood processing   Pulp and paper   Total

NOTE: The changes in value added are the changes in real value (i.e. adjusted for inflation).
SOURCE: FAO, 2008b.

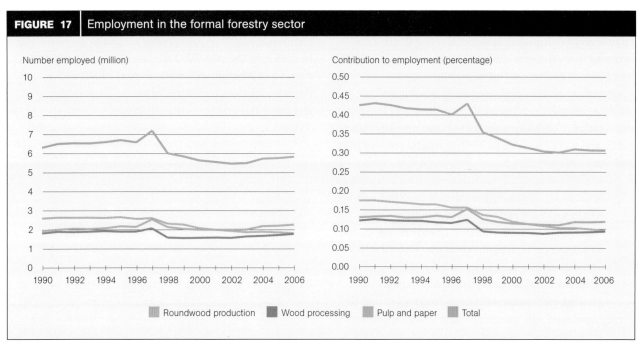

FIGURE 17 | Employment in the formal forestry sector

Number employed (million)

Contribution to employment (percentage)

Roundwood production   Wood processing   Pulp and paper   Total

SOURCE: FAO, 2008b.

Reducing the region's high rate of deforestation and forest degradation holds potential for mitigating climate change; much hope is pinned on the future of REDD initiatives under discussion in the context of the United Nations Framework Convention on Climate Change (UNFCCC).

Water scarcity is critical in some countries (especially Australia, China, India, Mongolia and Pakistan), affecting key sectors including agriculture and industry. The continued growth of most economies will depend on a sustained supply of freshwater. Public funding

for watershed management has received considerable attention, but market approaches are also being adopted, although most are still in the pilot stage of implementation (Dillaha et al., 2007).

Tourism in general, and ecotourism in particular, is one of the fastest-growing sectors in Asia and the Pacific, especially in view of the rapid growth in incomes. Most countries have developed national policies and strategies to promote ecotourism for its potential to revitalize local economies and protect and manage rural landscapes, including forests (Box 9). The main challenges arising from the growing demand for ecotourism are preventing environmental degradation and enhancing the income accruing to local communities, thus providing them with incentives to protect and manage natural assets.

Provision of most of the necessary environmental services depends on arresting deforestation and forest degradation. Considering the three broad development paths, the overall outlook for environmental services is as follows:

- In the postindustrial societies – with well-developed institutions, declining pressure on land and a strong will to maintain environmental quality – environmental protection has already received, and will continue to receive, substantial attention.
- The situation in the emerging industrial economies will be more varied. Although a growing environmentally conscious segment of the population will spearhead environmental protection initiatives, continued pressures of industrialization and the needs of marginalized people will strain the environment, particularly in countries with high population densities.
- In low-income forest-rich countries – which will need to cater to the demand for wood products, energy and industrial raw material from the rapidly growing economies, and to the demand for land from the expanding agricultural population – environmental protection is unlikely to receive much attention. These societies are less likely to be willing or able to pay for improving or maintaining environmental services.

SUMMARY

Considering the great diversity of the region, a varied scenario is expected to unfold. While forest area will stabilize and increase in most of the developed countries and some of the emerging economies, most of the low- and middle-income forest-rich countries will witness continuing decline owing to expansion of agriculture (including the production of biofuel feedstock). Both traditional woodfuel and emerging bioenergy options will pose enormous land-use challenges. The rapid industrialization of the emerging economies will create great demand for primary commodities, which is likely to result in forest conversion in the remaining countries.

Demand for wood products will continue to rise in line with the growth in population and income. While the region is at the forefront of plantation forestry, its dependence on wood from other regions will continue in the foreseeable future. Overall, the region – especially some of the most populous countries – faces severe land and water constraints that may limit the scope for self-sufficiency in wood products.

The demand for forest environmental services will increase as incomes rise. Conservation involving local communities will receive greater emphasis. It remains to be seen how the post-2012 climate change arrangements evolve and whether initiatives such as REDD will actually provide sufficient incentives to refrain from forest clearance and other unsustainable uses.

Europe

urope, consisting of 48 countries and areas (Figure 18), accounts for about 17 percent of global land area but has one-quarter of the world's forest resources, approximately 1 billion hectares, of which 81 percent is in the Russian Federation (Figure 19). Europe has a long tradition of multiple-use forest management with substantial emphasis on the provision of social and environmental services.

DRIVERS OF CHANGE

Demographics

Europe's population is projected to decline from 731 million in 2006 to 715 million in 2020 (Figure 20). This decline, together with the ageing of the population, will have important direct and indirect implications for forests and forestry. Declining labour supply will necessitate continued efforts to develop labour-saving technologies

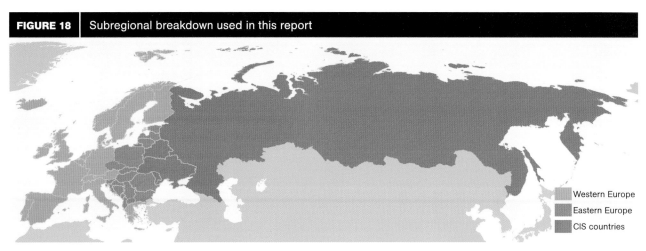

| FIGURE 18 | Subregional breakdown used in this report |

Western Europe
Eastern Europe
CIS countries

NOTE: See Annex Table 1 for list of countries and areas by subregion.

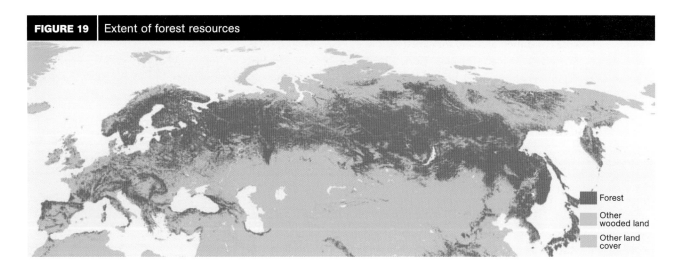

| FIGURE 19 | Extent of forest resources |

Forest
Other wooded land
Other land cover

FIGURE 20 | Population

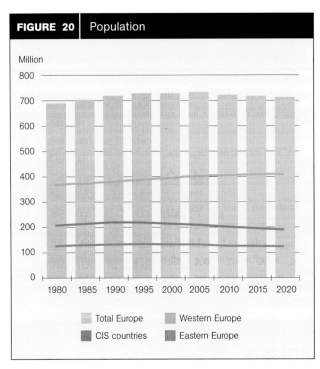

Million

1980	1985	1990	1995	2000	2005	2010	2015	2020

Total Europe   Western Europe
CIS countries   Eastern Europe

SOURCE: UN, 2008a.

and will encourage increased immigration and shifting production to low-wage economies. Immigration within the region is unlikely to last as wage rates converge.

In Europe, households are becoming smaller, and their number is expected to be 20 percent higher in 2030 than in 2005 – implying continued demand for construction timber, furniture and other wood products (EEA, 2005).

Within the region, population density falls in a gradient from southwest to northeast; most forests are in the less densely populated northern countries. The Russian Federation has only 9 people per square kilometre.

Western Europe is a highly urbanized subregion; more than 75 percent of its population is in urban areas. In some countries, urbanization exceeds 90 percent. However, there may be increased movement to rural areas (particularly mountains and coasts) as the population ages and the quality of life in urban centres declines, and this could increase pressure on forests (EEA, 2005). An

increase in the number of healthy and affluent retired people is likely to raise the demand for tourism, potentially in forests.

In general, Eastern European countries and the countries of the Commonwealth of Independent States (CIS) are relatively less urbanized, but political and economic changes are accelerating their pace of urbanization.

Economy

Despite differences among countries, Europe as a whole is characterized by relative economic stability and high income. Per capita income exceeds US$10 000 in all Western European countries, and US$35 000 in several. In contrast, most CIS countries have per capita incomes below US$10 000. The European Union has strengthened the growth of competitive market economies through common policies and the free flow of investments, technology, labour and goods, including forest products.

Economic forecasts suggest that Eastern European countries and the Russian Federation will grow much faster than Western Europe, albeit from a lower base figure (Table 8). The share of agriculture in GDP and employment is very low in Western Europe and is also declining in Eastern European and CIS countries (FAO, 2005b) in view of the faster growth of their manufacturing and services sectors. Land-use conflicts are declining as a result.

High income is reflected in relatively high consumption of forest products and an increasing demand for a broader range of forest-derived goods and services, with a strong emphasis on quality.

TABLE 8
GDP growth projections, 2000–2015

Subregion	GDP growth (%)
CIS countries	4.9
Eastern Europe	4.4
Western Europe	2.9
Total Europe	**3.4**

SOURCES: Based on UN, 2008b; World Bank, 2007a.

Policies and institutions

Europe has a strong overall political and institutional environment and favourable investment climate. Well-developed political systems have helped in establishing equilibrium between globalization and localization. Civil-society organizations are well developed, and public, private and civil-society organizations generally meet on a level playing field. Forest policies are largely developed through consultative processes.

The enlargement of the European Union and the growing role of the European Parliament in developing common strategies in critical areas have fostered political and institutional strengthening for many countries in the region. The main challenge for the European Union is to balance the different aspirations of its member countries within a common economic and political framework.

Forestry is a relatively minor economic activity in most European countries, hence the impact of policies in other sectors (agriculture, energy, industry, environment and trade) on the forest sector, or the contribution that the forest sector could make to the others, is not always taken into consideration.

Regional initiatives such as the Ministerial Conference on the Protection of Forests in Europe (MCPFE) and the European Commission's European Forestry Strategy provide effective coordination in forestry.

Science and technology

Europe is advanced in science and technology development, a large share of which is directly focused on Europe's most important source of income: high-technology manufacturing. Most Western European countries have a research and development (R&D) outlay of more than 2 percent of GDP (European Commission, 2007). Although the share of agriculture and forestry in the R&D budget is low, these fields benefit from technology developments in other sectors, especially in terms of improving industry practices and enhancing labour productivity. Remote sensing, information and communications technology and improved processing technologies have all benefited the forest sector. Future technological changes in the forest sector will be driven by:

- growing concern about climate change;
- the need to improve energy efficiency and reduce capital intensity;
- the desire for more sustainable forest management and more efficient use of forest resources, including recycling, reuse and conversion into bioenegy;
- focus on customer satisfaction and high-quality niche markets (Houllier *et al.*, 2005).

The European forest products sector will need to develop a new range of high-value-added products to meet the increasing demand for "green materials" and "green energy", to confront increasing competition from alternative materials and electronic media, and to compete with countries that have lower raw-material, energy and labour costs (CEI-Bois, CEPF and CEPI, 2005).

OVERALL SCENARIO

Although there are differences among the subregions, Europe generally presents a favourable situation in terms of social and economic development. Diminishing demographic pressures, moderate economic growth, well-developed political and institutional arrangements, growing concern for protection of the environment and especially for climate change, and high investments in science and technology are facilitating the transition to a knowledge-based postindustrial "green" economy built on the sustainable and equitable use of resources. This transition will take place at different speeds in the various countries.

Where there is strong political commitment to invest in green technologies and strengthen knowledge and skills, the transition will be rapid. However, in countries with lower incomes, environment and sustainability issues will be a low priority and the transition to a postindustrial society will be slower. In many cases, industries will move to countries where production costs are low (and environmental regulations are lax). Investments will continue to focus on improving competitiveness in the traditional sense, and unsustainable use of forests could continue.

OUTLOOK
Forest area

Europe has a relatively high proportion of its land area under forest (second only to Latin America and the Caribbean), which has consistently increased in recent years (Table 9). Growing stock per hectare is slightly lower than the global average but is high in some Western European countries (e.g. Austria and Switzerland) and in Eastern Europe, where, until recently, harvesting has been modest and silvicultural practices have favoured high stock accumulation.

The distinction between natural and planted forests is less clear for Europe than for other regions because much of the original forest cover was removed hundreds of years ago. Much of the region's increasing forest area reflects natural expansion of forests into former agricultural land and the establishment of semi-natural planted forests using native species.

Continued transition to a postindustrial society is expected to reduce pressure on forests, especially in Western Europe. Declining population, low land

TABLE 9
Forest area: extent and change

Subregion	Area (1 000 ha)			Annual change (1 000 ha)		Annual change rate (%)	
	1990	2000	2005	1990–2000	2000–2005	1990–2000	2000–2005
CIS countries	825 919	826 953	826 588	103	−73	0.01	−0.01
Eastern Europe	41 583	42 290	43 042	71	150	0.17	0.35
Western Europe	121 818	128 848	131 763	703	583	0.56	0.45
Total Europe	**989 320**	**998 091**	**1 001 394**	**877**	**661**	**0.09**	**0.07**
World	4 077 291	3 988 610	3 952 025	−8 870	−7 320	−0.22	−0.18

NOTE: Data presented are subject to rounding.
SOURCE: FAO, 2006a.

dependence, high income, concern for protection of the environment and a well-developed policy and institutional framework all favour further expansion of forest area. Almost all European countries have laws that make forest clearance and conversion to other land uses extremely difficult. In addition, fiscal support is provided for forestry under the European Agricultural Fund for Rural Development, encouraging significant expansion in tree planting. Thus, the forest area is likely to increase as the extent of land under agriculture decreases.

The major threats to forest resources in Europe are environmental (fires, pest outbreaks and storms); some of these could increase with climate change. Although the long-term impacts of climate change on forests are uncertain, many recent catastrophic events have been attributed to it. Considerable increases are projected in the extent and frequency of fires, for example in the Iberian Peninsula and in the Russian Federation (EEA, 2007).

Forest management
Forest management is influenced by the ownership structure. In Western Europe, 70 percent of forests are privately owned, often by individuals or families. In Eastern Europe, large parts of state forests were returned to their former owners in the 1990s, which increased the proportion of forests under private ownership (UNECE, MCPFE and FAO, 2007). Fragmented ownership among many smallholders raises the complexity and costs of forest management. In many countries, the private sector has responded by forming strong private forest owners' associations and cooperatives. In CIS countries, all forest is state owned.

Fellings in Europe have been lower than the growth in forest resources and have actually declined over several decades. In the future, the ratio of fellings to increment is expected to increase as more wood is harvested to supply the wood industry, as well as reflecting the impact of fast-growing demand for wood as a source of renewable energy.

In most countries, forest management is highly regulated with strict enforcement. State forest organizations play a leading role in forest management

as they have significant financial and technical resources. Western European countries tend to adopt intensive high-technology management involving improved planting stock, investments in soil improvements and mechanized harvesting. In Eastern Europe and the CIS subregion, where labour is cheaper, lower-cost management tends to be adopted with few inputs, long rotations and natural regeneration. Many absentee owners and smallholders also adopt this form of management.

A third form of management is traditional multipurpose management, whether carried out by the state (high-intensity multipurpose management) or in small, family-owned forests and farm forests to provide a range of non-wood benefits to their owners or local people. Forests managed in this way have suffered most in terms of economic viability with changing market conditions, i.e.:
- increased global competition resulting in lower product prices and reduced ability of industry to pay for wood and fibre;
- lower roundwood prices owing to the rapid increase in supply following forest restitution in Eastern Europe.

Balancing the economic forces of markets and the growing public demand for environmental and social services of forests will remain the major challenge. High labour costs and the complexity of managing many small fragmented forests make it difficult to meet the high forest management standards, reducing the economic viability of forest management in many countries, especially in Western Europe. There could be a shift towards production of smaller-sized wood grown on shorter rotations.

However, recent increases in demand for wood energy and higher prices could bring about a major shift from a wood-surplus to a wood-deficit situation.

Wood products: production, consumption and trade
Europe produces large amounts of a wide variety of wood products, is a major participant in international trade and has relatively high consumption (Table 10). The region accounted for almost one-third of global production in 2006 and roughly half of global wood products exports.

Western Europe has a major competitive advantage in the production of highly processed products such as reconstituted panels and high-quality paper. Its environmental concerns are reflected in, among others, its status as a major producer and consumer of certified wood products and its high rate of use of recovered fibre. In addition, governments and the private sector are promoting wood products and "green building" for their environmental friendliness.

The Russian Federation accounts for most of the forest industry in the CIS subregion. With its vast forest resources, low labour costs and technically skilled workforce, it has immense potential to regain its former position as a major global producer of wood products (Box 10).

Prior to 1990, Eastern European and CIS countries accounted for nearly half of Europe's sawnwood production. Political changes in the 1990s led to a drastic decline in their production and consumption of sawnwood. With the transition to a market economy, Eastern Europe shifted to production of more processed products such

as wood-based panels. Sawnwood production has begun to recover since 2000, but despite predicted growth of 1.7 percent from 2005 to 2020, it will still be less in 2020 than it was in 1990. Consumption is expected to remain relatively flat.

Production and consumption of wood-based panels are roughly equal and are expected to grow faster than those of sawnwood, 2.4 percent annually from 2005 to 2020, because of developments in reconstituted panel technology and substitution of panels for sawnwood.

High growth rates in paper and paperboard production are expected to continue, but with significant subregional differences (Figure 21). Europe's competitive advantage in paper production is based on close high-demand markets, availability of a large quantity of recovered paper and, in particular, technological sophistication for production of high-quality paper. The somewhat lesser competitive advantage of the CIS countries is based on abundant availability of pulpwood. Higher export tariffs in the Russian Federation (Box 10) are likely to stimulate increased pulpwood production in other European

TABLE 10

Production and consumption of wood products

Year	Industrial roundwood (million m³)		Sawnwood (million m³)		Wood-based panels (million m³)		Paper and paperboard (million tonnes)	
	Production	Consumption	Production	Consumption	Production	Consumption	Production	Consumption
2000	483	473	130	121	61	59	100	90
2005	513	494	136	121	73	70	111	101
2010	578	543	147	131	82	79	128	115
2020	707	647	175	151	104	99	164	147

SOURCE: FAO, 2008c.

| **BOX 10** | Recent developments in the forest industry in the Russian Federation |

Roundwood production in the Russian Federation was about 150 million cubic metres in 2005, amounting to one-third of all production in Europe and 10 percent of global production. However, this is still only about half the 1990 level. About one-third of the production was exported in 2005, accounting for 40 percent of global trade. The main importers of industrial roundwood from the Russian Federation are China, Finland and Japan.

Recent policies to stimulate domestic forestry include:
- the Forest Code (2007), which encourages private-sector participation in forestry (including through flexible forest lease arrangements) and decentralization of forest management;
- dramatically increased industrial roundwood export tariffs by 2009 (with a temporary two-year exemption for birch pulpwood) (see table).

However, the new policies do not address the scarcity of capital in forestry – caused by the perception of the Russian Federation as a high-risk country for investment and by the concentration of investments in the high-performing oil and gas sector. Thus, despite the advantages of abundant wood, low domestic wood prices (stumpage) and cheap skilled labour, it is unclear whether the policies will have the intended effect.

Year	Russian industrial roundwood export tariffs	
	(€/m³)	(US$/m³)
1996	4	5
2007	10	14
2008	15	23
2009	50	74

SOURCE: A. Whiteman, unpublished, 2008.

countries, particularly in northern Europe. In Eastern Europe, consumption of paper and paperboard is expected to outpace production, leading to increased imports. In contrast, paper and paperboard consumption in Western Europe is expected to remain flat because of substitution by electronic media.

Exports are high across almost all product sectors (Table 11). However, Europe's share in furniture has declined with the rise of Asia's furniture industry. Much of Europe's wood products trade (including roundwood) is within Europe, between Europe and North America and increasingly with Asia.

Europe is also one of the largest investors in the forest sector in emerging markets, particularly the pulp and paper sector in Asia and Latin America, where European companies benefit from matching their technological, marketing and managerial skills with the low labour costs, rapidly expanding planted forests and growing demand.

Over time, the differences in forestry development between Eastern and Western Europe are likely to diminish. Western Europe will remain focused on the production of highly processed wood products, supported by a high-technology approach to forest management, while wood production is expected to increase in Eastern Europe.

Woodfuel

The use of wood for energy in Europe became relatively minor after the Second World War. However, since the mid-1990s, the region (particularly the European Union) has introduced policies to increase the share of renewable energy in total energy consumption to combat climate change, meet Kyoto Protocol targets and address concerns about rising fossil fuel prices and energy security (Box 11).

These policies, together with market changes, have stimulated an increasing demand for wood as an energy source, and particularly for wood pellets as a substitute for oil in small-scale heating and electricity production (Box 12). In addition, within 5–10 years, the technology to produce liquid biofuels from wood could begin to be adopted on a commercial scale, which would increase woodfuel demand.

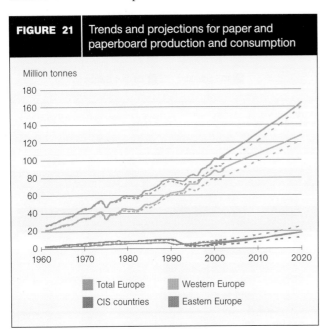

| FIGURE 21 | Trends and projections for paper and paperboard production and consumption |

NOTE: Solid lines represent production and dashed lines represent apparent consumption.
SOURCE: UNECE and FAO, 2005.

| BOX 11 | European Commission measures to promote renewable energy |

- Renewables Directive (2001): sets a target for electricity production from renewable sources of 22.1 percent by 2010
- Biofuels Directive (2003): sets an indicative target for consumption of liquid biofuels of at least 5.75 percent by 2010
- Draft Proposal for Climate Action (to come into force in 2010 if accepted): sets an objective of 20 percent of total energy from renewable sources by 2020 and a minimum target of 10 percent for market share of biofuels by 2020

TABLE 11

Exports as percentage of production and imports as percentage of consumption, 2006

Subregion	Industrial roundwood		Sawnwood		Wood-based panels		Paper and paperboard	
	Exports	Imports	Exports	Imports	Exports	Imports	Exports	Imports
	(%)							
CIS countries	34	1	68	3	27	22	35	28
Eastern Europe	14	8	49	27	45	44	59	67
Western Europe	9	19	46	46	51	48	67	61
Total Europe	**18**	**13**	**51**	**40**	**46**	**43**	**64**	**59**
World	**8**	**8**	**32**	**32**	**32**	**32**	**32**	**32**

SOURCE: FAO, 2008a.

The European Forest Sector Outlook Study (EFSOS) projected woodfuel consumption to 2020 (UNECE and FAO, 2005). However, new projections (Figure 22) are approximately three times higher for Eastern Europe and five times higher for Western Europe than the EFSOS figures, which were based on traditional woodfuel use mostly by households in rural areas (and were underestimated because of the paucity of reliable national statistics).

Fellings, thinnings and prunings, recovered wood products, residues from harvesting and processing and biomass from outside forests are all used for energy production. Wood used for energy needs to be fully taken into account in wood balance estimates; Table 12 suggests that when this is done, demand outweighs supply.

Non-wood forest products

Although not a major activity in Europe, the collection of NWFPs is a common form of recreation. Key commercial products include Christmas trees, game meat, cork, mushrooms (including truffles), honey, nuts and berries (Figure 23). Most of these have limited but well-established (and sometimes highly profitable) markets. Two recent developments include a decline in the viability of cork production (because of substitutes) and increased interest in food from forests as part of the growing consumer demand for organic products.

As with wood, European producers and forest managers have continuously adapted their practices to take advantage of the changing market conditions. For example, cork producers have improved marketing and introduced stricter quality controls, standards and certification to compete against substitutes. Producers

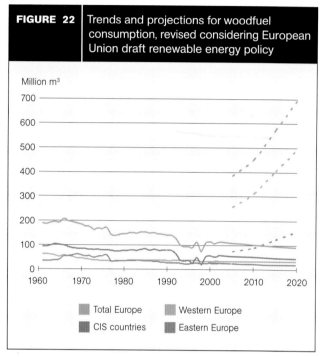

FIGURE 22 Trends and projections for woodfuel consumption, revised considering European Union draft renewable energy policy

NOTE: Solid lines are EFSOS projections and dashed lines represent revised projections.
SOURCES: Becker et al., 2007; UNECE and FAO, 2005.

BOX 12 | Emerging demand for wood pellets

Since wood pellets emerged in the 1970s as an alternative fuel source, their production and consumption have increased steadily, and new developments in manufacturing technologies have improved their quality. The availability of raw material, competitive prices and diversified energy policies favour the development of the wood pellet industry in Europe. In 2006, the overall production of almost 300 pellet plants in the European Union reached nearly 4.5 million tonnes. Sweden is the world leader in terms of wood pellet production. Sawdust-based pellet production has considerable potential in Brazil and the Russian Federation.

Consumption is also rising for both heating and electricity production (see figure). Globally, wood pellet markets display exponential growth, with new markets opening up in many areas, including Canada and Eastern Europe, and with potential in Asia and Latin America.

Future growth will depend on improved local logistics, a reduction in the cost of pellet stoves and supportive policies.

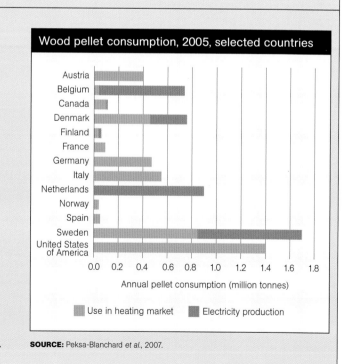

Wood pellet consumption, 2005, selected countries

SOURCE: Peksa-Blanchard et al., 2007.

TABLE 12

Components of wood supply and consumption, European Union and European Free Trade Association[a] countries, 2005

Supply	Million m³	% of total
From forest		
Industrial roundwood	397	51
Fuelwood	85	11
Bark	25	3
Logging residues	23	3
Woody biomass outside the forest	20	3
Co-products		
Chips, particles and wood residues	118	15
Pulp production co-products	70	9
Post-consumer recovered wood	29	4
Processed woodfuel industry	7	1
Total	**775**	**100**

Use	Million m³	% of total
Material		
Sawmill industry	217	26
Panel industry	88	11
Pulp industry	155	19
Pellets, briquettes, etc.	7	1
Other physical utilization	14	2
Energy		
Power and heat	49	6
Industrial internal	65	8
Private households	92	11
Undifferentiated energy use	135	16
Total	**822**	**100**

[a] Iceland, Liechtenstein, Norway and Switzerland.
NOTE: Data presented are subject to rounding.
SOURCE: Adapted from Mantau et al., 2008.

FIGURE 23	Marketed non-wood forest products in Europe, percentage of total value

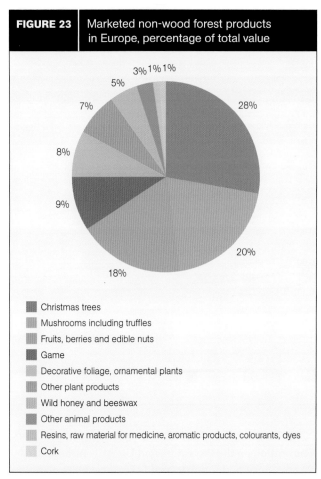

- Christmas trees
- Mushrooms including truffles
- Fruits, berries and edible nuts
- Game
- Decorative foliage, ornamental plants
- Other plant products
- Wild honey and beeswax
- Other animal products
- Resins, raw material for medicine, aromatic products, colourants, dyes
- Cork

NOTE: Based on available information.
SOURCE: UNECE, MCPFE and FAO, 2007.

of forest food products in Eastern Europe have taken advantage of low labour costs to serve the niche market for organic foods. In Western Europe, forest managers are earning income from NWFPs, for example through permits for recreational collection of mushrooms or sale of Christmas trees.

Contribution of forestry to income and employment

After the drastic decline that accompanied the political and economic changes in the early 1990s, gross value added by the forest sector recovered somewhat towards the middle of the decade, but it has continued to diminish since 2000 (Figure 24). Most of the decline has come from the pulp and paper subsector.

Employment in the forest sector has also fallen in absolute and relative terms (Figure 25).

Environmental services of forests

High levels of education and access to information contribute to great concern for protection of the environment in Europe, and high incomes contribute to willingness to pay for environmental services. Land use is highly regulated and forest clearance is virtually prohibited in most of the region, particularly in Western Europe.

Combating climate change is the most important environmental concern. In addition to having an expanding role in providing biomass for renewable energy (see section on woodfuel above), Europe's forests are also valued as a carbon sink. For Europe as a whole, land use, land-use change and forestry reduce net emissions by almost 6 percent, and forests probably account for almost all of this reduction (Table 13). The contribution is particularly high in Eastern Europe, where increment is high and emissions from other sectors are low. Europe has also pioneered market approaches for emission trading.

Protected areas in Europe expanded from 195 million hectares in 1990 to 234 million hectares in 2007 (UN, 2008c). There are multiple initiatives for conserving biodiversity in the region (Box 13), although most measures that maintain biodiversity in forests are not specifically earmarked as such. Management practices increasingly emphasize protecting biodiversity through

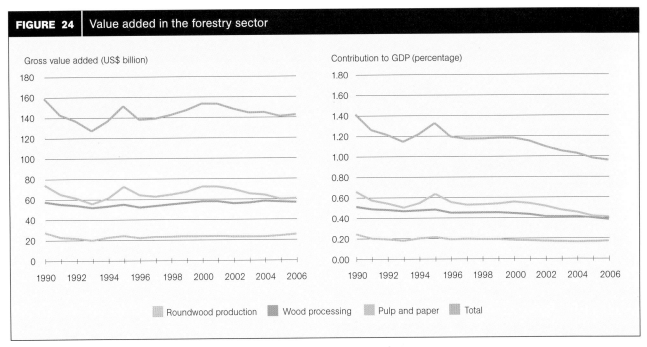

FIGURE 24 | Value added in the forestry sector

Gross value added (US$ billion)

Contribution to GDP (percentage)

Roundwood production ▪ Wood processing ▪ Pulp and paper ▪ Total

NOTE: The changes in value added are the changes in real value (i.e. adjusted for inflation).
SOURCE: FAO, 2008b.

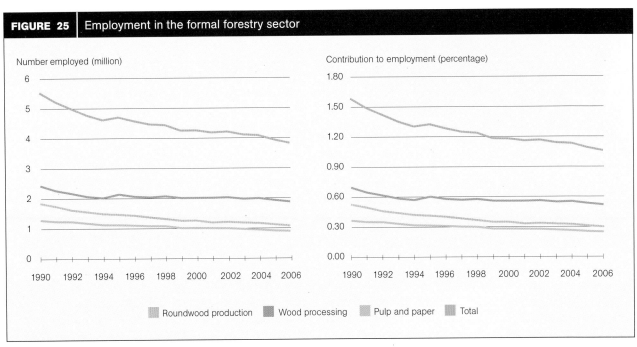

FIGURE 25 | Employment in the formal forestry sector

Number employed (million)

Contribution to employment (percentage)

Roundwood production ▪ Wood processing ▪ Pulp and paper ▪ Total

SOURCE: FAO, 2008b.

natural regeneration, mixed forests, leaving dead wood in forests and protecting small "key habitats" in managed forests (UNECE, MCPFE and FAO, 2007). Growing emphasis on "close to nature silviculture" (UNECE, FAO and ILO, 2003) will help to conserve biodiversity in most managed forests.

Integrated management of upland watersheds and the linkages between forests and water are receiving increasing attention in the region. In 2006, the United Nations Economic Commission for Europe (UNECE)

Convention on the Protection and Use of Transboundary Watercourses and International Lakes (also known as the Water Convention) endorsed the concept of PES, including the conservation and development of forest cover. In 2007, MCPFE adopted a resolution on forests and water that addresses, among others, policy coordination and economic valuation of water-related forest services. FAO, UNECE, MCPFE and the European Commission highlighted forest–water linkages at European Forest Week in October 2008 (UNECE and FAO, 2008).

TABLE 13

Impact of land use, land-use change and forestry (LULUCF) on net emissions of greenhouse gases, 2005 (as reported to UNFCCC)

Subregion	Total greenhouse gas emissions (Mt CO$_2$e)		Contribution of LULUCF to net emissions		Contribution of wood energy to net emissions			Contribution of wood energy and LULUCF	
	Without LULUCF	With LULUCF	Total (Mt CO$_2$e)	As % of emissions without LULUCF	Consumption of woodfuel (million m^3)	Avoided fossil fuel emissions (Mt CO$_2$e)	As % of emissions without LULUCF	Total (Mt CO$_2$e)	As % of emissions without LULUCF
CIS countries	2 627	2 700	+73	+2.8	56	−22	−0.9	+51	+1.9
Eastern Europe	1 298	1 082	−216	−16.7	76	−30	−2.3	−247	−19.0
Western Europe	4 306	3 966	−340	−7.9	257	−103	−2.4	−443	−10.3
Total Europe	**8 231**	**7 748**	**−484**	**−5.9**	**389**	**−156**	**−1.9**	**−639**	**−7.8**

NOTES: Mt CO$_2$e = megatonnes CO$_2$ equivalent. Data presented are subject to rounding.
SOURCE: Mantau *et al.*, 2008.

BOX 13	Ecological networks in Europe

- Pan-European Ecological Network (PEEN): aims to enhance ecological connectivity across Europe by promoting synergies between nature policies, land-use planning and rural and urban development
- Natura 2000: a network of Special Protection Areas for birds and Special Areas of Conservation for other species and habitats, established by European Union legislation and involving up to 20 percent of the European Union's land area
- Emerald Network: initiated under the Convention on the Conservation of European Wildlife and Natural Habitats (also known as the Bern Convention), extends a common approach to the designation and management of protected areas to European countries (non-European Union) not covered by Natura 2000 as well as to Africa

SOURCE: EEA, 2007.

More than 90 percent of European forests are open to public access and the area of forest available for recreation is increasing. Ecotourism is popular. While the demand for forests as recreation areas is expected to increase, the nature of the demand is expected to change, influenced by demographic and income changes (Bell *et al.*, 2007).

The transition to a green economy requires strong demand, and willingness to pay, for forest environmental services. Europe's high income, increasing area of forests and growing focus on multiple-use management with more emphasis on environmental values suggest positive movement in this direction. Multifunctional forestry with a greater focus on the provision of environmental services requires a strengthening of cross-sectoral policy coordination; this remains a challenge in some areas.

SUMMARY

Forest resources in Europe are likely to continue expanding. Fellings will probably remain below increment, and the provision of environmental services will continue as a primary concern, especially in Western Europe. Rules and regulations in this regard will make wood production less competitive in comparison with other regions.

Forest management will continue to serve a wide variety of demands. Economic viability is likely to remain a challenge, especially for small-scale forest owners, but the increased demand for woodfuel could change this.

While the forest industry, especially in Western Europe, may continue to lose competitiveness against other regions in labour-intensive segments, it is likely to retain leadership in the production of technologically advanced products, with much of the forest industry shifting to the production of "green" products.

Within the region, the differences in forestry between Eastern and Western Europe are likely to diminish as Eastern Europe catches up. The impacts of recent developments in the Russian Federation and in promoting wood energy are difficult to predict, and at present are mainly addressed for the short term.

Latin America and the Caribbean

atin America and the Caribbean, consisting of 47 countries and areas (Figure 26), accounts for 22 percent of the global forest area, 14 percent of the global land area and 7 percent of the world's population (Figure 27). The region contains the world's largest contiguous block of tropical moist forest – the Amazon Basin.

DRIVERS OF CHANGE

Demographics

The population in the region is projected to increase from more than 450 million in 2005 to 540 million by 2020 (Figure 28). Population density is low, averaging 25 people per square kilometre in 2006, although this figure is dominated by South America, with 21 people per square kilometre. In Central America and in the Caribbean, there are 79 and 179 people per square kilometre, respectively. Population density in the region is expected to exceed 30 people per square kilometre by 2020 (UN, 2008d). The most populous country in the region, Brazil, which accounts for 41 percent of the region's population, has a density of only 22 people per square kilometre, while at the other extreme Bermuda has 1 280 people per square kilometre.

The urban population makes up 78 percent of the total population and is expected to reach 83 percent by 2020. Fourteen percent of the urban population resides in one of four megacities (of 10 million inhabitants or more). Many South American countries encourage settlement in frontier areas to counter urbanization and attendant social and economic problems (UN, 2008d).

Economy

Almost all countries in the region are in the middle-income bracket and growing rapidly, although growth is uneven in many countries (Figure 29). While per capita income is high in comparison with other developing regions (with several countries exceeding US$5 000 per year), income remains unevenly distributed. In some countries, the richest one-tenth of the population receive nearly 50 percent of the total income and the poorest one-tenth less than 2 percent.

Globalization will continue to drive change in the region. Important influences are bilateral and multilateral trade agreements and the growing investment and trade linkages with the emerging Asian economies, particularly China and India. Given the export-driven industrialization policies, continued global demand for agriculture, livestock, forest products and, increasingly, biofuel will intensify the pressure on forests.

Increasing emphasis on export-led growth implies that economic performance will be influenced by changes in global markets and competitiveness. Liberalization policies have led to substantial increases in investments, boosting

| **FIGURE 26** | Subregional breakdown used in this report |

Caribbean
Central America
South America

NOTE: See Annex Table 1 for list of countries and areas by subregion.

growth rates. Forecasts suggest that economic growth will remain high (World Bank, 2007a; UN, 2008b), but changes in global markets and growing competition from emerging Asian economies could alter the trend.

With rapid development of the manufacturing and services sectors, agriculture's share in GDP (only 7 percent in 2005) and employment has decreased in most countries. However, while the viability of small-scale agriculture has declined with import liberalization, large-scale export-focused commercial agriculture, including livestock, has expanded impressively (e.g. soybeans, biofuel crops, meat, fruits, vegetables and cut flowers) (World Bank, 2007b)

| FIGURE 27 | Extent of forest resources |

- Forest
- Other wooded land
- Other land cover

| FIGURE 28 | Population |

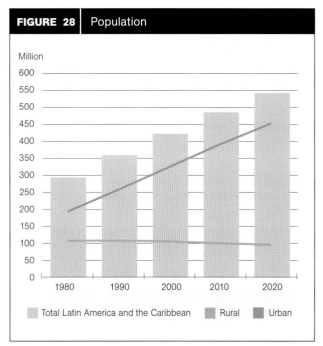

Total Latin America and the Caribbean   Rural   Urban

SOURCE: UN, 2008a.

| FIGURE 29 | Gross domestic product |

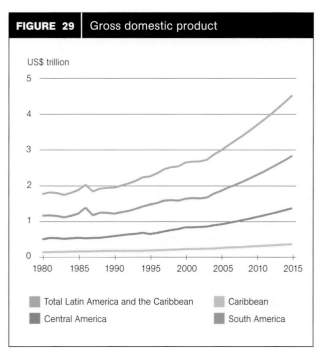

Total Latin America and the Caribbean   Caribbean
Central America   South America

SOURCES: Based on UN, 2008b; World Bank, 2007a.

and is responsible for most of the region's deforestation (Figure 30).

While the region's resource-rich countries are increasingly linked to the rest of the world as producers of industrial goods and primary commodities, others with high population densities and limited resources are witnessing a different kind of globalization, largely linked to the provision of services (e.g. tourism).

Policies and institutions

In the past two decades, democratically elected governments have largely replaced authoritarian regimes in the region. Political changes have not significantly affected broad policies, which commonly pursue growth with varying emphasis on redistribution.

A pluralistic institutional environment has emerged with government, the private sector and civil-society organizations having an important role in forest resource management. Of particular interest to forestry are:

- decentralization, particularly recognition of the rights of local and indigenous communities to manage natural resources (Box 14);
- greater private investment in managing natural and planted forests;
- substantial incentives contributing to the rapid growth of planted forests, including low-interest loans and tax breaks;

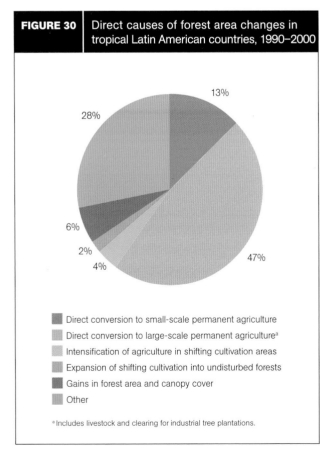

| FIGURE 30 | Direct causes of forest area changes in tropical Latin American countries, 1990–2000 |

13%
28%
6%
2%
4%
47%

- Direct conversion to small-scale permanent agriculture
- Direct conversion to large-scale permanent agriculture[a]
- Intensification of agriculture in shifting cultivation areas
- Expansion of shifting cultivation into undisturbed forests
- Gains in forest area and canopy cover
- Other

[a] Includes livestock and clearing for industrial tree plantations.

SOURCE: FAO, 2001.

BOX 14 | Indigenous community forest ownership

In the past two decades, some countries have granted legal ownership of forest to indigenous communities: Bolivia, 12 million hectares; Brazil, 103 million hectares; Colombia, 27 million hectares; Ecuador, 4.5 million hectares; and Guyana, 1.4 million hectares of land including forests.

While ownership gives the communities secure rights to sustainable use of forest resources, disputes over ownership (sometimes violent) and lack of enforcement of rules and regulations have allowed illegal occupation and logging in vast areas of these forests.

SOURCE: ITTO, 2006.

- the growing role of local, national and international civil-society organizations in forest issues, including rights for indigenous communities, forest certification and combating illegal logging and forest clearance – with special focus on Amazon forests owing to their global significance for biodiversity conservation and climate change mitigation.

Science and technology

Although most countries in the region spend less than 0.5 percent of GDP on R&D, investments in R&D are increasing. Brazil, the regional leader, spends 1 percent of its GDP on R&D (still below the international average of 2–3 percent) and has established a legal framework for investing in science and technology (the Innovation Act of 2004). Funding arrangements for science and technology have improved, with particular efforts to link research institutions with industry (de Brito Cruz and de Mello, 2006).

Research areas of particular interest to forestry in the region include information and communication technologies, remote-sensing technology to monitor forest area changes, productivity-enhancing technologies for planted forests, precision logging systems and biofuel technologies (especially cellulosic biofuel). Brazil is already a global leader in sugar-based ethanol production.

OVERALL SCENARIO

Countries in the region are likely to follow two broad patterns of development:

- Natural-resource-dependent economic development: Countries with low population densities and substantial forest resources will take advantage of the increasing global demand for food, fuel and fibre. The main challenge will be determining the trade-offs between the different options. While there will be significant efforts to conserve resources, the emphasis on immediate economic gains through large-scale expansion of production of food, fuel and fibre is likely to dominate in the short term.

• Shift from dependence on natural resources: More densely populated and relatively resource-poor countries will emphasize human-resource-based development. Urbanization and emerging alternative sources of income (including remittances from workers abroad) could help to reduce land pressures. The economic viability of small farms will continue to decline, resulting in less-intensive cultivation or even abandonment. Increasing income will also result in greater willingness to improve the environment.

OUTLOOK

Forest area

In countries with relatively high forest cover and in the early stages of industrialization, forests are highly vulnerable. Between 1990 and 2005, the region lost almost 64 million hectares, or 7 percent, of its forest area (Table 14). The region accounted for more than one-third of annual global forest area loss from 2000 to 2005.

All South American countries registered a net forest loss between 2000 and 2005 except Chile and Uruguay, which had positive trends because of large-scale industrial plantation programmes. With the increasing global demand for food, fuel and fibre, those forest-rich countries in South America that remain dependent on natural resources will continue to lose forests to large-scale industrial agriculture and cattle ranching as long as these remain competitive. New planted forests for industrial uses, especially in Argentina, Uruguay and potentially Colombia, may partially offset the loss of natural forests, although not in ecological terms.

In most Central American countries, net forest loss declined from 2000 to 2005 in comparison with the previous decade, with Costa Rica achieving a net increase in forest area. However, in percentage terms, Central

America has had one of the highest rates of forest loss of any subregion in the world, exceeding 1 percent per year from 2000 to 2005. This rate is expected to decline as small-scale agriculture becomes uneconomic, with abandonment of marginal farmlands, increasing opportunities for alternative sources of income and growing urbanization. Several countries in the subregion will witness stabilization and recovery in their forest area.

The Caribbean registered a small increase in forest area between 2000 and 2005, mainly in Cuba. Trade liberalization, which has made traditional agricultural exports such as sugar and bananas uncompetitive, is resulting in abandonment of agricultural land and reversion to secondary forest (Eckelmann, 2005). Furthermore, greater emphasis is being given to protecting the natural environment in support of the growing tourism industry (Box 15). Thus, forest area is expected to remain stable or to expand in most Caribbean countries.

Forest management

Although the role of natural forests in wood production is declining with the rise of plantation forestry, they remain an important source of timber in some countries. Natural production forests are largely managed through long-term

---

BOX 15	Tourism in the Caribbean

The Caribbean accounts for 5.1 percent of total global demand for tourism. Tourism contributes 16.5 percent to the subregion's gross domestic product and its contribution is predicted to remain stable until at least 2014. Tourism directly employs 15 percent of the total population and indirectly supports close to half of the population. Given the dependence on coastal zones for attracting visitors, global warming and natural disasters such as hurricanes are increasingly drawing attention to environmental conservation issues.

SOURCE: Griffin, 2007.

TABLE 14
Forest area: extent and change

Subregion	Area (1 000 ha)			Annual change (1 000 ha)		Annual change rate (%)	
	1990	2000	2005	1990–2000	2000–2005	1990–2000	2000–2005
Caribbean	5 350	5 706	5 974	36	54	0.65	0.92
Central America	27 639	23 837	22 411	−380	−285	−1.47	−1.23
South America	890 818	852 796	831 540	−3 802	−4 251	−0.44	−0.50
Total Latin America and the Caribbean	**923 807**	**882 339**	**859 925**	**−4 147**	**−4 483**	**−0.46**	**−0.51**
World	**4 077 291**	**3 988 610**	**3 952 025**	**−8 868**	**−7 317**	**−0.22**	**−0.18**

NOTE: Data presented are subject to rounding.
SOURCE: FAO, 2006a.

private concessions of up to 200 000 ha in Bolivia, Guyana and Suriname; medium-sized concessions in Guatemala, Peru and the Bolivarian Republic of Venezuela; and small-scale concessions in Colombia, Ecuador, Honduras and Trinidad and Tobago (ITTO, 2006). In Brazil, nearly all production has been in private forests, but the Law on the Management of Public Forests for Sustainable Production approved in 2005, and now beginning to be put into practice, opens up national forests in the Amazon for logging concessions; the intention is to encourage sustainable management and help to avoid illegal occupation and logging (Box 16).

Selective logging is the primary focus of most concession management in the region, with little attention to postharvest silviculture and unregulated harvesting leading to degradation. Obstacles to sustainable management of the region's natural forests for wood production include:

- scarce adoption of reduced-impact logging because of weak incentives;
- limited area of forests certified (Box 17) because of the high costs and absence of a price premium, especially with the availability of low-priced illegally procured timber;

BOX 16 | Brazilian forest concessions

The Law on the Management of Public Forests for Sustainable Production outlines the allocation of timber concessions in Brazil's federal forests. Salient features of the law include:

- creation of the Brazilian Forest Service;
- establishment of the National Forest Development Fund;
- allocation of forest concessions through a transparent and open bidding process;
- preference given to non-profit organizations, communities and non-governmental organizations;
- allocation of 20 percent of concession revenues to the Brazilian Forest Service and the Brazilian Institute of Environment and Renewable Natural Resources.

The emphasis is on safeguarding environmental, social and economic values. Bids are judged on price only after demonstration that operations will cause the least environmental impact, generate the largest direct social benefits and add the most value to products and services in the concession area.

Private logging concessions are expected to cover 13 million hectares within the next decade, eventually expanding to about 50 million hectares.

SOURCES: Schulze, Grogan and Vidal, 2007; Tomaselli and Sarre, 2005.

BOX 17 | Forest certification

In 2007, Latin America and the Caribbean had about 12 million hectares of certified forests, or about 4 percent of all certified forests in the world. Although the certified area represented only 1.2 percent of the region's forests, this was a significant increase from 0.4 percent in 2002. Almost 80 percent was certified by the Forest Stewardship Council, and the rest under national systems: CERFLOR (Brazil) and CERTFOR (Chile), which is affiliated with the Programme for the Endorsement of Forest Certification. Brazil's CERFLOR has separate standards for natural and planted forests.

SOURCE: ITTO, 2008.

- ownership disputes from overlapping land tenure and illegitimate titles encouraging illegal logging and land conversion, especially in the Amazon;
- diseconomies of scale for small community-managed concessions, especially those remote from markets;
- preponderance of the informal sector (especially illegal logging and wood-processing units).

Considering the conflicting demands, multiple-use management of natural forests continues to be a complex challenge. The difficulties will discourage long-term private investments, and most logging will continue to be done by short-term investors.

Latin America and the Caribbean has about 12.5 million hectares of planted forests. This is only 5 percent of the global planted forest area (FAO, 2006b), but the region is emerging as a leader in high-productivity plantation forestry. Argentina, Brazil, Chile and Uruguay account for about 78 percent of the planted forests in the region. Plantation development, driven by the private sector, is supported by favourable government policies and financial incentives. These include partial reimbursement of costs, tax breaks and low-interest loans for small owners (Box 18). These factors have made South America a destination for investments by both regional and global pulp and paper producers and recently by North American investors, including timber investment management organizations (TIMOs).

Key features of plantation forestry in the region include:
- investment in productivity-enhancing technologies, especially clonal propagation, achieving productivity of more than 50 m³ per hectare per year in some cases;
- use of intensively managed short-rotation species such as *Eucalyptus* spp., radiata pine (*Pinus radiata*), loblolly pine (*Pinus taeda*) and southern yellow pine (*Pinus elliottii*);
- integration of plantation management with wood processing, especially pulp and paper and panel production.

Current projections suggest an increase in the area of planted forests in the region from 12.5 million hectares in 2006 to 17.3 million hectares in 2020 (see Box 31 on page 63).

Availability of suitable land and a favourable investment climate will enable the region (primarily South America) to maintain its competitive advantage in plantation forestry. As a high proportion of production is geared to global markets, the future of plantation forestry will depend on global demand, especially for pulp and paper, panel products and biofuel feedstock. A possible increase in transportation costs could be a major concern, especially if wood products are destined to meet the demand from the emerging Asian economies.

Wood products: production, consumption and trade

Industrial wood production, while not significant in Central America or the Caribbean, is increasing rapidly in South America, especially because of plantation

investment in the Southern Cone. The region's share of global industrial roundwood production rose from 7 percent in 1990 to 10 percent in 2006. Production of key products, in particular pulp and paper, has grown since 1990 and the trend is likely to continue considering the high investments in plantations and processing (Table 15).

Domestic consumption of wood products is essentially stable (Figure 31). Increasing income could boost consumption in some countries, and housing programmes will boost domestic timber consumption despite competition from substitutes used in construction. However, the domestic market for most products is expected to remain small except in Brazil.

Most production is exported. The net export value of all products exceeded US$7 billion in 2005. However, the net export value has declined recently (Figure 32) as a result of the appreciation of South American currencies against the United States dollar and because of increasing competition from China, especially in furniture and panel products.

Export promotion programmes will continue to encourage the production of paper and packaging. The region's share of the global market in pulp and paper products will increase, especially with continuing disinvestments in Europe and North America and the relocation of wood products industries to regions that

---

BOX 18	Incentives for forest plantations in Chile and Uruguay

In Chile, government policies in place for some decades to promote planted forests and private investments have resulted in a strong diversified forest industry and a plantation area of more than 2 million hectares. The national development strategy promotes financial incentives for industrial forest plantations. Legal instruments define subsidies and regulate logging, favouring small and medium-sized landholdings and plantations in degraded areas. The forestry sector now accounts for about 20 percent of Chile's exports and 4 percent of its gross domestic product.

In Uruguay, the government has supported planted forests since 1987 by granting tax benefits when they are established in Forestry Priority Areas (extending over 2.5 million to 3 million hectares). Inexpensive flat terrain and favourable climate and soil provide ideal conditions. In 2005, Uruguay had 0.8 million hectares of planted forests and an annual planting rate of 50 000 ha.

SOURCE: PwC, 2007a.

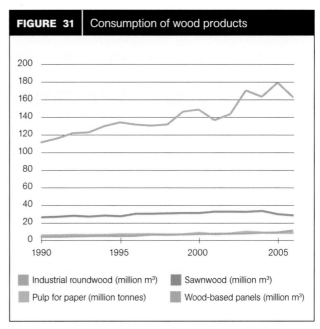

FIGURE 31	Consumption of wood products

Industrial roundwood (million m³) Sawnwood (million m³)
Pulp for paper (million tonnes) Wood-based panels (million m³)

SOURCE: FAO, 2008a.

TABLE 15
Production and consumption of wood products

Year	Industrial roundwood (million m³)		Sawnwood (million m³)		Wood-based panels (million m³)		Paper and paperboard (million tonnes)	
	Production	Consumption	Production	Consumption	Production	Consumption	Production	Consumption
2005	168	166	39	32	13	9	14	16
2020	184	181	50	42	21	12	21	24
2030	192	189	60	50	29	15	27	31

SOURCE: FAO, 2008c.

have competitive advantages. South America's advantages include a stable investment climate, low population density, favourable conditions for tree growth and significant technical capacity. Consequently, South America has some of the lowest wood fibre costs in the world (PwC, 2007b).

Woodfuel

Household woodfuel use is declining in South America (mainly because of urbanization and increased use of fossil fuel and biofuels), steady in the Caribbean and rising in Central America. Overall, woodfuel production in the region has been growing gradually over the past ten years. This trend is expected to continue (Figure 33), mainly owing to industrial charcoal use in Brazil (Box 19). Future demand will also depend on the supply of fossil fuels and developments in renewable energy technologies.

Non-wood forest products

Most NWFPs in the region are for local subsistence use, although some are sold in national and international markets as ingredients for health and beauty care products and medicines. Brazil nuts (*Bertholletia excelsa*) are an important source of income for indigenous groups in Bolivia, Brazil and Peru and are also the most important commercial NWFP; the supply chain provides direct employment for 15 000 people. Brazil nuts constitute 45 percent of Bolivia's forest-related exports (more than that of all wood products) and contribute more than

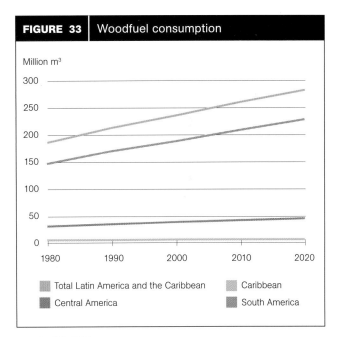

FIGURE 33 | Woodfuel consumption

Million m³

Legend:
- Total Latin America and the Caribbean
- Central America
- Caribbean
- South America

SOURCE: FAO, 2003b.

BOX 19 | Charcoal for iron and steel

Aside from spearheading the most extensive global programme to introduce biofuels (ethanol) into its energy matrix, Brazil also consumes large quantities of charcoal in its iron and steel industry – an estimated 8.3 million tonnes in 2006. Iron and steel companies and others involved in supplying charcoal to the industry own about 1.2 million hectares of forest plantations, which produced almost 10 million tonnes of charcoal in 2005.

SOURCE: UN, 2008f.

US$70 million per year to the national economy (CIFOR, 2008a).

To reduce conflicts between NWFP-dependent indigenous communities and loggers and ranchers in the Amazon, Brazil has established extractive reserves exclusively for the collection of NWFPs. This model, which grants long-term rights in public forests to groups engaged in sustainable activities, is spreading through the region. Initiatives supported by civil-society organizations and governments have improved NWFP collection, value addition and marketing, with the support of certification and fair trade organizations.

As economies grow and urbanize and more lucrative income-earning opportunities become available, dependence on NWFPs for subsistence is expected to decline. Processing and marketing of products that are already well known will improve. Local value chains will largely be replaced by national and global chains, often assisted through fair trade initiatives and organic labelling.

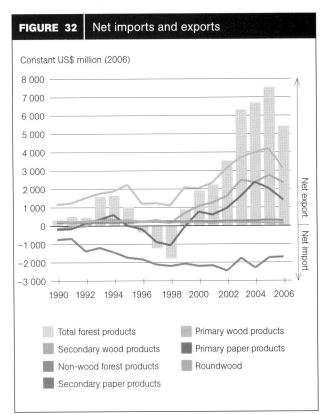

FIGURE 32 | Net imports and exports

Constant US$ million (2006)

Net export | Net import

Legend:
- Total forest products
- Secondary wood products
- Non-wood forest products
- Secondary paper products
- Primary wood products
- Primary paper products
- Roundwood

SOURCES: FAO, 2008a; UN, 2008e.

Contribution of forestry to income and employment

Since 1990, the contribution of forestry to GDP has increased from US$30 billion to $40 billion (Figure 34). Most of the increase in gross value added is from roundwood production. Value added in wood processing and pulp and paper production has remained stable, but the latter is expected to change with the increasing investments in pulp and paper capacity. Employment in the forestry sector has also increased (Figure 35). In comparison with other regions, the share of forestry in total value added and employment has remained relatively stable.

Environmental services of forests

The impact of deforestation on the region's provision of global and regional environmental services (biodiversity, water regulation, climate change mitigation and nature-based tourism) is drawing particular attention. While non-market interventions (through policies and legislation) have been the primary means for environmental conservation, the region is a leader in adopting market-driven approaches, especially PES schemes. In most cases, these are not strictly market-driven approaches, but primarily government-managed schemes using tax revenues to pay landowners, with no direct linkage

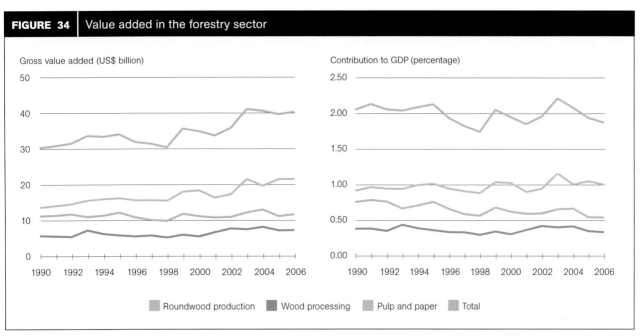

FIGURE 34 Value added in the forestry sector

NOTE: The changes in value added are the changes in real value (i.e. adjusted for inflation).
SOURCE: FAO, 2008b.

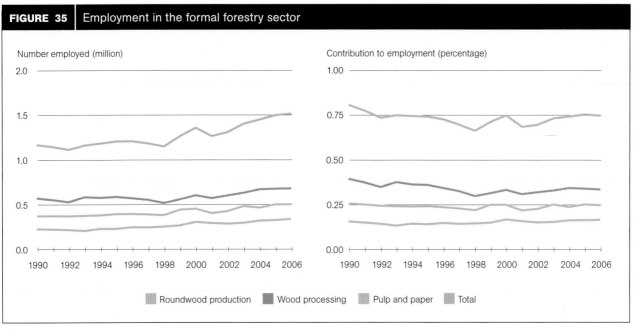

FIGURE 35 Employment in the formal forestry sector

SOURCE: FAO, 2008b.

between providers and buyers of environmental services (Kaimowitz, 2007).

Brazil, Colombia, Ecuador and Peru rank among the world's ten most biodiverse countries, while the eastern slope of the Andes is the most biologically diverse area in the world. Ten countries each have more than 1 000 different tree species. However, the region also leads the world in the number of tree species considered endangered or vulnerable to extinction (FAO, 2006a). Forty percent of the plant life in the Caribbean is found nowhere else (USAID, 2006).

Establishment of protected areas has been core to environmental conservation in the region. Between 1990 and 2007, the extent of protected areas increased from 213 million to 451 million hectares (24 percent of the world's protected areas) (UN, 2008c). However, many governments have limited human and financial capacity to enforce conservation measures. Conservation often comes into conflict with mining, oil extraction, agriculture and logging, particularly where property rights are ill-defined.

The outlook for maintaining and improving watershed services also depends on land-use changes. It looks bleak considering the high rate of deforestation. Water scarcity is particularly acute in the Andes and in some of the Caribbean islands. The region has been a pioneer in implementing payment for watershed services. In most cases, the schemes are managed by intermediary organizations, often government agencies responsible for managing irrigation and domestic water supply facilities, which channel funds from water users to landowners. There is potential to improve and scale up some of the initiatives. However, wider adoption will depend on overcoming some obstacles. These include ill-defined property rights; farmers' fears that their resources will be expropriated; distrust of privatization of water supply; and inadequate information on the technical linkage between upstream land use and downstream benefits (Dillaha *et al.*, 2007).

With its high deforestation rate, the region has great potential for reducing greenhouse gas emissions through slowing deforestation and degradation.

Ecotourism is an important income generator in several countries, especially in the Caribbean. The highly diverse ecosystems make the region one of the most popular ecotourism destinations. For example, Costa Rica has taken advantage of its natural attractions and made ecotourism the backbone of its economy. Ecuador earns more than US$100 million per year from nature-based tourism in the Galapagos Islands. Easier access and higher incomes could result in continued growth of ecotourism in the region – although concern about carbon footprints and further ecosystem degradation may begin to deter ecotourists. Concern is growing about threats to biodiversity from increased numbers of visitors. Managing tourism sustainably and enhancing its benefits to the poor will remain the major challenges.

PES systems, including those proposed under the REDD initiative, will surely gain momentum. However, it remains to be seen whether they can bring about significant changes in the behaviour of those responsible for forest clearance. PES appears to be particularly effective where the opportunity cost of land use is low.

SUMMARY

The outlook for forests and forestry in Latin America and the Caribbean will be influenced by the pace of diversification of the economies and changes in land dependence (FAO, 2006c).

In most Central American and Caribbean countries, population densities are high; as urbanization increases, there is a significant shift away from agriculture and related activities, especially as smallholder agriculture becomes less remunerative. Tourism and remittances from migrant workers are becoming important sources of income. Agriculture-related forest clearance is declining and some cleared areas will revert to forest, as is already evident.

Although population density is low in South America, high food and fuel prices will favour continued forest clearance for increased production of livestock and agricultural crops for food, feed and biofuel to meet global demand – especially as South American economies increase linkages with emerging Asian economies.

Planted forests will spread, promoted by private investments and continuing global demand for wood products, especially from the emerging Asian economies. However, the accelerated plantation rate will not offset continuing deforestation.

In short, the pace of deforestation in South America is unlikely to decline in the near future. Heavily forested countries that are taking advantage of the expanding global demand for primary products and are pursuing a path of rapid economic development will find it extremely difficult to slow the rate of forest conversion. Provision of global public goods – for example carbon credits – may help to some extent. However, an effective mechanism for providing adequate incentives to refrain from forest clearance has yet to be developed.

North America

The North America region, consisting of 3 countries and 2 areas (Figure 36), has 7 percent of the world's population, 16 percent of its land area and 17 percent of its forest area (677 million hectares). About one-third of the region's land area is forested (Figure 37). The highly varied climate conditions create great diversity in forest ecosystems, ranging from humid tropical to boreal. Some of the world's most productive forests are found in this region.

DRIVERS OF CHANGE

Demographics

North America's population is expected to increase from about 441 million in 2006 to 500 million in 2020 (Figure 38). The annual population growth rate, much influenced by immigration, is 0.9 percent (but declining) in Canada and 1 percent in both Mexico and the United States of America (hereafter "United States").

The region has a low population density of about 21 people per square kilometre, ranging from fewer than 4 in Canada to 54 in Mexico. Nearly 80 percent of the population is urban, and urbanization is expected to continue, with the greatest growth in Mexico. Despite considerable demand for outdoor recreational activities,

there is concern that urbanization is disconnecting people from nature. The "More Kids in the Woods" project implemented by the United States Forest Service is an attempt to reverse the situation (ARC, 2007).

The ageing population in Canada and the United States is reducing the size of the labour pool and, hence, the availability of workers for forestry. Immigration is helping to overcome labour shortages to some extent.

Economy

The region accounted for 32 percent of global GDP in 2006, although this share is declining. From 2000 to 2006, GDP growth was about 3 percent. GDP is forecast to increase from US$15 trillion in 2006 to more than US$20 trillion in 2020. The United States accounts for more than 80 percent of regional GDP (Figure 39).

Poverty and income disparity are important issues in the region. About 35 percent of the rural population in Mexico (World Bank, 2004) and about 12 percent in the United States (USDA, 2004) were estimated to be below the poverty level in 2002.

As a result of the shift from an agricultural to an industrial economy in the twentieth century, agriculture now accounts for less than 1 percent of GDP in Canada and the United States. This transition also contributed to the

FIGURE 36 | Subregional breakdown used in this report

- Canada
- Mexico
- United States of America

FIGURE 37 | Extent of forest resources

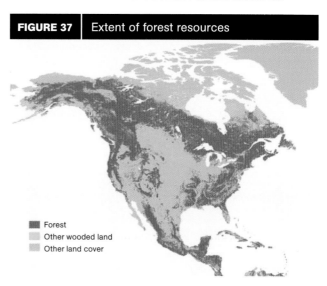

- Forest
- Other wooded land
- Other land cover

NOTE: See Annex Table 1 for list of countries and areas by subregion.

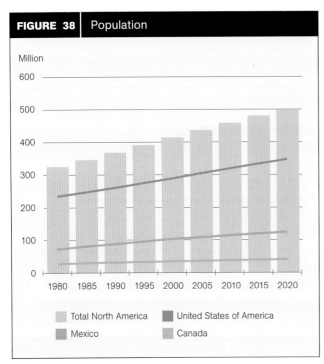

FIGURE 38 | Population

Million

600
500
400
300
200
100
0

1980 1985 1990 1995 2000 2005 2010 2015 2020

Total North America   United States of America
Mexico   Canada

SOURCE: UN, 2008a.

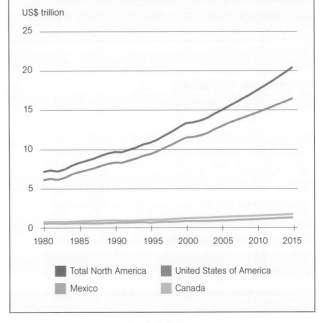

FIGURE 39 | Gross domestic product

US$ trillion

25
20
15
10
5
0

1980 1985 1990 1995 2000 2005 2010 2015

Total North America   United States of America
Mexico   Canada

SOURCES: Based on UN, 2008b; World Bank, 2007a.

stabilization of forest area (MacCleery, 1992). Mexico is still in the transformation phase; agriculture's share in GDP declined from 13 percent in 1970 to 8 percent in 1990 and 4 percent in 2006 (World Bank, 2007a). However, it remains important for employment in the country (19 percent of employment in 2004) (FAO, 2005b). Although commercial agriculture has grown rapidly, subsistence agriculture also remains prominent, particularly under the system of *ejidos* (communally held lands) and other traditional community arrangements. Agriculture-related deforestation remains high.

North America is one of the most actively globalizing regions, with a high level of inflow and outflow of capital, labour and technology. Substantial natural and human resources and a high level of innovation enhance its global competitiveness. However, increasing competition from low-cost producers (especially China) and the inclination to offshore or outsource production in order to remain competitive are transforming some sectors, including forestry (Box 20).

BOX 20 | Impact of globalization on the forestry sector in the United States of America

- About one in six pulp and paper mills has closed since the mid-1990s.
- One-third of pulp and paper mill jobs have disappeared since the early 1990s because of consolidation, cost-cutting and productivity improvements.
- The number of large softwood sawmills declined from 850 to 700 in 2004 alone.
- Sales of imported wooden household furniture, primarily from China, increased from about 20 to 50 percent in the 1990s and they have continued to expand.

SOURCE: Ince *et al.*, 2007.

Mexico's export-focused industrialization is being challenged by competition from rapidly industrializing Asian economies in both domestic and global markets – and particularly in United States markets, which absorb more than 80 percent of all Mexican exports.

Since 2006, the United States has been experiencing an economic slowdown, which is also affecting the Canadian and Mexican economies because of the interdependence of the countries in the region. A related slump in the construction sector has influenced the demand for wood products (discussed below). Import liberalization under the North American Free Trade Ageement (NAFTA) has had mixed impacts; while exports have increased, wages and living conditions have declined. Expansion of large-scale commercial agriculture and displacement of small farmers have accelerated poverty-related deforestation (Audley *et al.*, 2004).

Policies and institutions

Public institutions are well developed and have continuously adapted to the larger economic and social changes (MacCleery, 2008). Stakeholder consultation helps to incorporate diverse perceptions in public decision-making.

The private sector has a pivotal role in all economic activities, although this is a recent trend for Mexico and several key nationalized industries remain. Large corporations have been leaders in innovation. Industry is becoming more consolidated through mergers and acquisitions.

Community-based organizations have an important role in natural resource management and have helped indigenous communities, especially in Canada, to cement their rights to hold land and to manage natural resources

(Box 21). Mexico has a long history of community management of natural resources under the *ejidos*. Policies promoting privatization and changes in the rural economy (particularly in agriculture and migration) are enabling the *ejidos* to benefit from opportunities for processing and trading wood and other forest products.

Civil-society organizations, especially in Canada and the United States, contribute to shaping policies and strategies in the forest sector and encourage social and environmental responsibility in the corporate sector. Civil action, together with industry consolidation and technological changes, has transformed the forest sector, especially in the western United States. Legal action initiated by civil-society organizations caused a radical reduction in timber supply from national forests in the 1990s. Such organizations are also growing in importance in Mexico.

Science and technology

Well-established institutions for science and technology and substantial public and private investments in research have enhanced competitiveness in all sectors, including forestry. In Mexico, investment-linked technology transfer has helped advance forestry (as well as agriculture), although many industries in Mexico, especially the smaller ones, still use old equipment and technology.

The forest industry has continuously improved processing technologies, enhancing productivity in order to withstand global competition. Especially during economic downturns, the industry tends to close plants that are less economically viable and invest in new plants with improved technologies.

High fossil fuel prices and concerns about energy security and climate change are stimulating investments in new energy technologies. The pulp and paper industry is diversifying into biorefining, producing a stream of products, including biofuels, electricity and chemicals (see Box 48 on page 93). Substantial research on producing cellulosic fuels is under way, focusing especially on efficient and cost-effective technologies for breaking down cellulose.

While the United States has long been a leader in science and technology, it is concerned about the possibility of losing this position as other regions (particularly Asia and Europe) accelerate investments in this area (Task Force on the Future of American Innovation, 2005). For example, the number of research scientists employed by the United States Forest Service has declined by about 75 percent in the past 30 years, with progressively more research being funded by the private sector (US Forest Service, personal communication, 2008).

BOX 21	Indigenous people and Canada's forests

- More than three-quarters of Canada's indigenous communities reside in forested areas.
- The forest products industry employs more than 17 000 indigenous people directly and indirectly, although many are still in lower-skilled, part-time and seasonal positions.
- The forest industry does business with more than 1 400 firms run by indigenous people.
- About 1 000 forestry operations are owned by indigenous people.

SOURCE: Natural Resources Canada, 2007a.

TABLE 16
Forest area: extent and change

Country/region	Area (1 000 ha)			Annual change (1 000 ha)		Annual change rate (%)	
	1990	2000	2005	1990–2000	2000–2005	1990–2000	2000–2005
Canada[a]	310 134	310 134	310 134	0	0	0	0
Mexico	69 016	65 540	64 238	−348	−260	−0.52	−0.40
United States of America	298 648	302 294	303 089	365	159	0.12	0.05
Total North America[b]	**677 801**	**677 971**	**677 464**	**17**	**−101**	**0**	**−0.01**
World	**4 077 291**	**3 988 610**	**3 952 025**	**−8 868**	**−7 317**	**−0.22**	**−0.18**

[a] Because data from previous inventories cannot be compared meaningfully, figures from the most recent inventory are given for all three reporting years (FAO, 2006a).
[b] Regional total includes Greenland and Saint Pierre and Miquelon.
NOTE: Date presented are subject to rounding.
SOURCE: FAO, 2006a.

OVERALL SCENARIO

North America has a generally favourable demographic, political, institutional and technological environment. However, the current economic slowdown in the United States and the larger global economic changes (especially the emergence of Asian economies) pose some uncertainty for the future outlook. If the downturn continues, reduced demand, low investments and declining incomes would lead to reduced consumer spending, loss of profitability and reduced public funding in most sectors, including forestry. Increasing competition and the tendency of the private sector to outsource or offshore production could spur protectionist measures, slowing global trade growth.

On the other hand, economic recovery in the United States (and by association the rest of the region) would boost demand for all products and investments in innovation, accelerating the transition to a knowledge-based economy. This scenario would provide opportunities for continued rapid industrial growth, modernization and poverty reduction in Mexico.

BOX 22	Mountain pine beetle infestation in British Columbia, Canada

Spreading through western Canada, the mountain pine beetle (*Dendroctonus ponderosae*) has infested 13 million hectares of pine forests (mostly lodgepole pine, *Pinus contorta*) and is expected to kill up to 80 percent of all pine stands in the Province of British Columbia. More than 530 million cubic metres of timber had been lost in British Columbia by 2007 and it is predicted that 1 billion cubic metres will be lost by 2018. The loss of trees is releasing more carbon than that from forest fires in spite of efforts to salvage the timber (which continues to store carbon).

The beetle is native to North America, but its range has spread northward and to higher elevations with milder winters. Temperatures below −40 °C on several consecutive nights will kill the larvae, but such cold spells have become rare.

SOURCES: Brown, 2008; Natural Resources Canada, 2007b.

OUTLOOK
Forest area

Forest cover in the region is stable. North America accounted for an estimated 2 percent of annual global deforestation from 2000 to 2005, although the rate of loss has been decreasing. Most of the loss was in Mexico, attributed mainly to agricultural expansion and unsustainable logging, while the United States reported a small net gain in forest area for the period (Table 16).

In the United States, forest area stabilized in the early twentieth century (MacCleery, 1992). A net loss of about 2 million hectares of forest is projected between 1997 and 2020; this estimate includes conversion of forest land to other uses, including urban and suburban development, as well as afforestation and natural reversion of abandoned crop and pasture land to forest (US Forest Service, 2008).

Change in forest area is not statistically significant in Canada. At even the highest estimates, it would take 40 years for Canada to lose 1 percent of its forest area (Canadian Council of Forest Ministers, 2006).

However, climate change may intensify threats to forest health. The intensity and frequency of forest fires have increased in both Canada and the United States, exacerbated by prolonged drought (attributed to climate change) and successful fire control programmes that have inadvertently increased the amount of combustible material. Climate change is similarly exacerbating pest infestations; in the west of Canada and the United States, the mountain pine beetle is causing particularly serious damage and tree loss (Box 22).

The situation in Mexico is more uncertain and depends on the pace of transition from an agrarian to an industrial society and the impact it may have on poverty and land dependence. Higher incomes could improve society's ability to invest in conservation and forest management (Comisión Nacional Forestal, Mexico, 2008). In recent years, the government has increased substantially the allocation to the forest sector, and this could have a positive impact on sustainable forest management. However, a reduction in economic growth could impede improvements.

Forest management

Differences in forest ownership explain much of the variation in forest management in the region.

In Canada, 92 percent of forests are publicly owned and managed to satisfy multiple needs (social, cultural, environmental and economic) in accordance with the National Forest Strategy, adopted with broad input in 2003 (FAO, 2006a). The Canadian Council of Forest Ministers has developed a national criteria and indicators framework for sustainable forest management at the provincial and local levels. Canada has the world's largest area of third-party certified forest (more than 134 million hectares). Annual harvest levels remain below the increment.

In the United States, private forests dominate in the east and public forests in the west. Overall, 58 percent of forests are private (FAO, 2006a). Wood production from public forests has been scaled down in response to the increasing demand for environmental services. More than 60 percent of wood production comes from non-industrial private lands and 30 percent from industry-managed forests. In the past decade, a major development in forest ownership has been the divestment of woodlands controlled by large forest companies. As a result, millions of hectares of forest land have passed into the hands of newly emerged TIMOs and real estate investment trusts (REITs) as well as families and others (see Box 41 on page 83). This ownership fragmentation increases the unit costs of management and may jeopardize its stability.

In Mexico, 8 500 *ejidos* or other community organizations own an estimated 59 percent of the forests (FAO, 2006a). The effectiveness of community forest management varies depending on the capacities and constraints of the communities and alternative land-use opportunities. In 2002, only about 28 percent of the forest-owning *ejidos* and communities carried out commercial harvesting activities (ITTO, 2005). Some *ejidos* engage in wood processing (e.g. sawnwood, furniture and floorings) and some have obtained certification from the Forest Stewardship Council (FSC) or SmartWood. Government compensation is available to communities willing to set forests aside for provision of environmental services rather than production.

If economic difficulties in the United States persist, forest management could suffer (Box 23), particularly in private forests as the pressures of responding to short-term economic changes may undermine owners' commitment to long-term sustainable forest management. If instead the economy should improve rapidly, the outlook for forestry would be much brighter, especially as the revival of the construction sector in the United States would stimulate demand for wood and, consequently, investment in management.

Wood products: production, consumption and trade

North America is the world's largest producer, consumer and exporter of wood products. In 2006, the region produced 38 percent of the world's industrial roundwood. This share has been generally stable since 1990, with wood production hovering around 600 million cubic metres per year (Figure 40).

| BOX 23 | Probable consequences if the economic downturn continues in the United States of America |

Canada
- Overall decline in wood production because of the drop in demand and closure of manufacturing plants (despite the market being flooded with softwood for several years because of the mountain pine beetle infestation)
- Reduced investment in forest management because of the shrinking market, allowing increases in fire and pest infestations, especially with climate change

Mexico
- Declining demand for timber from managed forests, and consequent decline in the ability of community organizations to manage forests
- Increased illegal logging as a result of loss of jobs in community enterprises and weakening of community control

- Expansion of subsistence cultivation and consequent deforestation and degradation

United States of America
- Slump in housing demand and consequent scaling down of production and employment in forest industries
- Significant reduction in investment in forest management by the private sector leading to further divestment and fragmentation of privately managed forests, which may eventually be converted to other land uses
- Decline in investment in public forests

Sawnwood production in North America increased from 128 million to 154 million cubic metres between 1990 and 2006, while global production declined. The regional increase largely reflects demand from the United States construction sector. However, the recent slump in this sector has reduced demand, although this may be temporary.

Production of wood-based panels rose from 44 million to 62 million cubic metres between 1990 and 2006 (with Canada accounting for most of the increase), but the region's relative share declined as global production doubled in the same period.

North America's share in global production of paper and paperboard also declined, from 39 percent in 1990 to 29 percent in 2006, largely because of the expansion of capacity in Asia and Latin America. This downward trend is unlikely to change in the coming years. Widespread use of electronic media is reducing the demand for paper, particularly in Canada and somewhat in the United States, although it is projected to increase in Mexico.

Long-term growth in net imports of wood products in the United States has been a consequence of rising demand from the construction sector (until recently) and declining domestic production. The United States has been a net importer since 1992, with the trade deficit reaching US$37 billion in 2005 (Figure 41). However, the recent slowdown in construction has improved the United States' wood products trade balance.

Canada remains a net exporter of wood products, with a trade surplus of about US$20 billion in 2006. However, exports have declined since 2005 with the

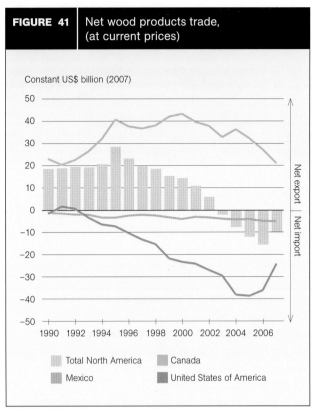

| FIGURE 41 | Net wood products trade, (at current prices) |

SOURCES: FAO, 2008b; UN, 2008e.

construction slump in the United States and also with the appreciation of the Canadian dollar against the United States dollar, which makes Canadian imports more expensive. This decline is forcing a scaling down of Canadian production. An important issue is whether the wood industry in Canada will be able to diversify and become less dependent on markets in the United States, which absorbed 78 percent of Canadian exports in 2006 (Natural Resources Canada, 2008a). In the short term, this may be especially challenging in view of the large supply increases expected from salvage operations in forests infested with mountain pine beetle in western Canada.

Mexico remains a net importer of wood products (with a trade gap of US$6 billion in 2007). The exception is secondary wood products (especially furniture), for which Mexico's exports, mainly to the United States, have reached US$1 billion in recent years. However, in 2007, Mexico's exports of secondary wood products declined and imports increased because of the economic situation in the United States and tightened competition from East Asian countries.

After having long been an attractive market, North America now presents considerable short- and medium-term uncertainties in terms of demand for wood products. Projections based on historical trends suggest a moderate increase in consumption of key products (Table 17) if the current decline is brief.

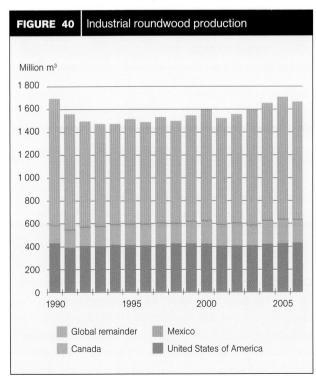

| FIGURE 40 | Industrial roundwood production |

SOURCE: FAO, 2008a.

TABLE 17

Production and consumption of wood products

Year	Industrial roundwood (million m³)		Sawnwood (million m³)		Wood-based panels (million m³)		Paper and paperboard (million tonnes)	
	Production	Consumption	Production	Consumption	Production	Consumption	Production	Consumption
1990	591	570	128	117	44	43	91	87
2005	625	620	156	158	59	70	109	106
2020	728	728	191	188	88	96	141	138
2030	806	808	219	211	110	115	169	165

Woodfuel

In 2005, woodfuel contributed about 3 percent of total energy consumption in the United States, about 4.5 percent in Canada and about 5 percent in Mexico (IEA, 2007). Woodfuel demand in Mexico has been declining because of urbanization and improved access to other energy sources (including fossil fuels), but household dependence on woodfuel remains high in some rural areas; the volume of wood extracted for fuel may be up to four times that of industrial timber production. Most woodfuel is harvested without a management scheme.

In Canada and the United States, the wood products industry leads in the use of energy from biomass, producing its own heat and electricity using cogeneration technology. The pulp and paper industry in Canada derives 57 percent of its energy from forest biomass.

Policy initiatives responding to escalating energy costs and climate change are expected to enhance the use of wood energy (Box 24). The demand for wood pellets for use in heating has increased significantly in recent years. The United States consumed the largest amount of wood pellets for this purpose in 2006, around 1.4 million tonnes (see Box 12 on page 28). In 2006, Canada and the United States produced about 1.5 million and 1 million tonnes of wood pellets, respectively, ranking second and third behind Sweden. Eventual commercial-scale cellulosic biofuel production could have important impacts on the forest sector.

Non-wood forest products

Rural communities in Mexico depend on NWFPs for subsistence and income, although their use is declining rapidly because of urbanization, changes in employment and availability of cheaper alternatives. NWFP harvesting in Canada and the United States typically takes place as part of forest recreation and cultural traditions, and it is increasing. Production of the few economically important NWFPs with long-established markets – notably maple syrup and Christmas trees – is highly commercialized. Both markets have been stable since 1994 and are expected to remain so. Canada accounts for 85 percent of the world's maple syrup production and the United States

BOX 24	Examples of policy initiatives to promote bioenergy

Canada
- Clean Air Agenda (2006): sets federal emission targets, allocates resources for the ecoENERGY for Renewable Power programme and promotes blended transportation fuel
- Regulatory Framework for Air Emissions: uses carbon credits to encourage renewable power production through cogeneration

Mexico
- Law for the Promotion and Development of Bioenergy (2008): aims to promote biomass energy without compromising food security

United States of America
- Energy Independence and Security Act (2007): sets targets for biofuel use (including wood-derived biofuels) to 2022 and sets a national fuel economy standard of 15 km per litre by 2020
- Biofuels Initiative (2006): aims to make cellulosic ethanol cost-competitive by 2012 and to replace 30 percent of current petrol consumption with biofuels by 2030

produces the rest. Canada produced 3.2 million Christmas trees in 2005 (Natural Resources Canada, 2008a).

Markets for herbal products, including forest medicinal plants, are expanding as society becomes increasingly health conscious. Large pharmaceutical companies are investing in the production and marketing of herbal plant products, which have become a multibillion-dollar industry in the United States (Alexander, Weigand and Blatner, 2002).

Contribution of forestry to income and employment

Overall, the gross value added by the region's forestry sector has increased from about US$130 billion in 1990 to US$148 billion in 2006 (Figure 42). Most of the increase is attributed to wood processing, while pulp and paper production has marginally declined. However, gross value added as a proportion of GDP has dropped from about 1.4 percent to less than 1 percent.

The number of people employed by the sector declined by about 140 000 between 1990 and 2006 (Figure 43), reflecting technological changes and improved productivity. Forestry employment accounted for about 0.8 percent of total employment in 2006.

As more forests are taken out of production, both gross value added and employment in the forestry sector are expected to decline.

Environmental services of forests

As income increases, society tends to assign greater importance to environmental conservation. Especially in Canada and the United States, a host of institutions – public, private, community and civil society – are involved in issues of climate change mitigation, biodiversity conservation and maintaining water supplies. These countries have a robust political and regulatory framework for environmental protection. Complex political processes have been developed to balance trade-offs between competing objectives and interests.

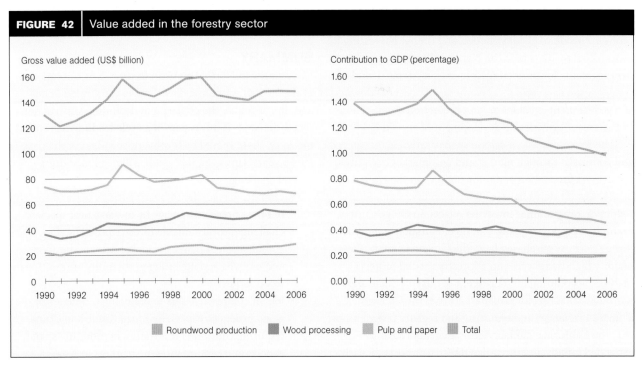

FIGURE 42 Value added in the forestry sector

Gross value added (US$ billion)

Contribution to GDP (percentage)

Roundwood production   Wood processing   Pulp and paper   Total

NOTE: The changes in value added are the changes in real value (i.e. adjusted for inflation).
SOURCE: FAO, 2008b.

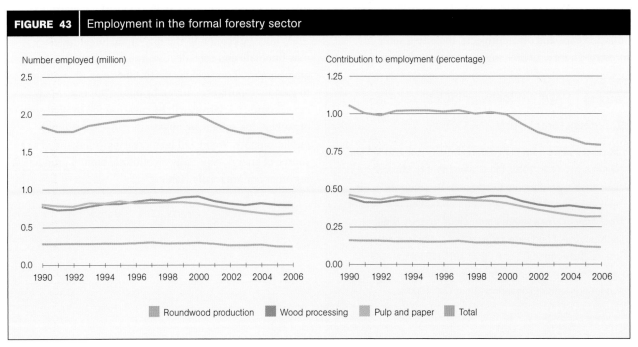

FIGURE 43 Employment in the formal forestry sector

Number employed (million)

Contribution to employment (percentage)

Roundwood production   Wood processing   Pulp and paper   Total

SOURCE: FAO, 2008b.

Mexico is one of the five most biologically diverse countries in the world, but continued dependence on land and consequent forest clearance are challenges to biodiversity protection in the country (Conservation International, 2005).

North America had 360 million hectares of protected areas in 2006, of which more than 70 percent was in the United States (UN, 2008c). A series of legislative and regulatory actions protect wildernesses and exclude large tracts of public land from logging or land-use changes; an example in the United States is the Roadless Areas Conservation Rule of 2001, which establishes prohibitions on road construction and timber harvesting in inventoried roadless areas on National Forest System lands. Arrangements for protecting unique ecosystems include conservation easements – agreements between landowners and government agencies or land protection organizations ("land trusts") restricting development of certain lands.

The role of forests in carbon sequestration is recognized and addressed through market and non-market initiatives involving afforestation and reforestation. In the United States, several states have initiated mandatory emission reduction programmes involving offsets. In Oregon, for example, new power plants can meet emission standards through offsets purchased from the Oregon Climate Trust, under which there were three forestry-related projects in 2008, accounting for 21 percent of offsets (Gorte and Ramseur, 2008). Voluntary markets (e.g. the Chicago Climate Exchange) and reporting and registry programmes (e.g. the California Climate Action Registry) have expanded rapidly and recognize forestry projects. In early 2008, three regional partnerships – Regional Greenhouse Gas Initiative, Western Climate Initiative and Midwestern Greenhouse Gas Reduction Accord – involved 23 states of the United States and 4 provinces of Canada in developing emission caps and offset projects, including some in forestry. These activities suggest continued growth of carbon markets and possibly an increasing role of forestry provided it is seen as an economically viable option.

Forests' role in water provision is important. Mexico has recently initiated a system for payment for water services (Box 25). Similar initiatives exist in Canada and the United States.

In Canada and the United States, outdoor recreation is a major use of forests and woodlands and has become an important source of income in many forested areas. In the United States, one in five leisure travellers visited national forests in 2006 (ARC, 2006).

SUMMARY

Uncertainty in North American forestry is a consequence of the current economic downturn in the United States and, in particular, the consequent declining construction sector demand. If this is part of a cycle leading to eventual recovery, there should be few major surprises in the next 10–15 years. However, the sector will need to address several challenges:

- climate change, the increasing frequency and severity of forest fires and damage by invasive pest species;
- challenges to sustainable forestry posed by the combination of increased global demand for food and biofuels and declining profitability of traditional wood industries;
- loss of competitiveness to emerging producers of wood products, especially Brazil, Chile and China, requiring continued innovation in order to expand exports and capture growing markets in Asia.

In Mexico, the rate of deforestation will continue to decline as urbanization continues and as increasing investments in reforestation and improved management practices result in more sustainable forest management.

While the economic viability of the forest industry may fluctuate and even decline, the provision of environmental services in North America will continue to gain in importance, driven by public interest. Many conservation initiatives will be spearheaded by civil-society organizations, which are able to mobilize substantial public support. Wood will be increasingly demanded as a source of energy, especially if cellulosic biofuel production becomes commercially viable.

BOX 25	Payment for hydrological services in Mexico

Mexico suffers both high deforestation rates and severe water scarcity. In 2003, the Government of Mexico launched a programme to compensate landowners for maintaining forests for watershed protection and aquifer recharge in areas where commercial forestry is not competitive. Funds are collected annually from water users. Between 2003 and 2006, US$110 million was allocated to landowners (both private and community) under agreements covering about 500 000 ha.

SOURCE: Muñoz-Piña *et al.*, 2006.

Western and Central Asia

Western and Central Asia, consisting of 25 countries and areas (Figure 44), is the least forested region in the world, with only 4 percent forest cover (1.1 percent of the global forest area) (Figure 45). A few countries account for most of the forest area; 19 countries have less than 10 percent forest cover. About 75 percent of the region is arid, with low biomass productivity. Vegetation ranges from desert scrub in Central Asia and the Arabian Peninsula to pockets of mangrove forests on the Persian Gulf coast and alpine meadows in Central Asia. In view of the low forest cover, trees outside forests, especially on farms and in other wooded land, have important productive and protective functions.

DRIVERS OF CHANGE
Demographics
Western and Central Asia's population is expected to increase from 371 million in 2006 to 479 million in 2020 (Figure 46). Population in the region is projected to grow at an annual rate of 2 percent between 2005 and 2020.

While Armenia, Azerbaijan, Georgia and Kazakhstan have low to negative growth rates, several countries – for example Afghanistan, Qatar, the Syrian Arab Republic, the United Arab Emirates and Yemen – have growth rates exceeding 2.5 percent. A high proportion of the population is less than 14 years old, implying considerable growth in the working-age population in the next two decades and a consequent need for more jobs, housing and amenities. Intensifying this need is the high rate of urbanization; in Western Asia, for example, 78 percent of the population is forecast to be urban by 2020. Urbanization is also increasing the demand for green spaces, bringing about important changes in forest policies (Amir and Rechtman, 2006).

Economy
Economic growth in the region has been robust in the past decade (IMF, 2008), largely because of the rising price of fossil fuels. Continued global demand will keep energy prices high, sustaining a high rate of income growth in the next decade and beyond (Figure 47). With the exception of a small number of non-fossil-fuel-producing countries, per

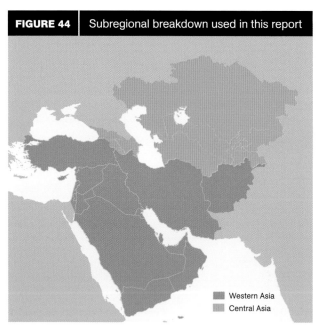

FIGURE 44 | Subregional breakdown used in this report

Western Asia
Central Asia

NOTE: See Annex Table 1 for list of countries and areas by subregion.

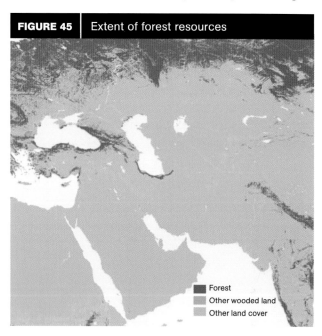

FIGURE 45 | Extent of forest resources

Forest
Other wooded land
Other land cover

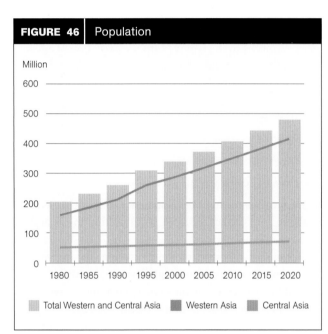

FIGURE 46 | Population

Million

SOURCE: UN, 2008a.

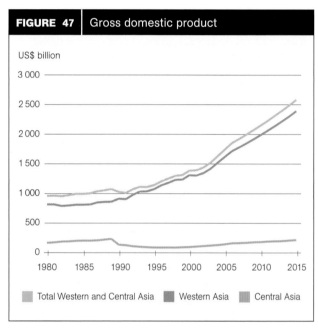

FIGURE 47 | Gross domestic product

US$ billion

SOURCES: Based on UN, 2008b; World Bank, 2007a.

capita incomes will continue to increase, although income distribution may remain skewed.

However, non-fossil-fuel-producing and less diversified economies, such as Afghanistan, Kyrgyzstan, Tajikistan and Yemen, face a number of challenges. While some have benefited from the spill-over effects of the high income of the wealthier fossil-fuel-producing countries (especially through employment, markets for products and tourism), poverty remains high in these countries, as does dependence on agriculture.

The region also has a number of high- and middle-income countries with highly diversified economies including agriculture, manufacturing and a vibrant services sector. For example, Israel is among the most innovative countries in agriculture and high-technology manufacturing.

Realizing that dependence on fossil fuels creates vulnerability, most countries that produce them are diversifying their economies through investment in agriculture, industries and the services sector, including tourism. The recent rise in food prices has encouraged

some of the Gulf Cooperation Council (GCC) countries to invest in agricultural projects in countries outside the region where land and water are more available.

Although agriculture and animal husbandry account for a declining share of GDP in correspondence with the expansion of other sectors such as fossil fuels and minerals, industries and services (FAO, 2007b), they remain vital for most countries, including those that have alternative sources of income. In some countries, for example Saudi Arabia, a reduction in subsidies for high-input agriculture in dry areas has resulted in a shift of agriculture to areas with a more favourable climate, including forested zones, resulting in forest clearance.

In most countries, livestock numbers have increased substantially, largely to cater to the increasing demand for meat. Higher incomes have enabled herders to transport livestock over long distances and to new grazing areas, and even to transport water. While traditional nomadic livestock management ensured sustainability of rangelands, the new practices and the increased numbers of animals have accelerated the degradation of forests and rangelands (FAO, 2008f). In some of the fossil-fuel-rich countries, former herders and farmers who have moved to urban areas employ migrant workers to take over their former occupation; hence, the pressure on forests and rangelands continues.

Policies and institutions

Policies and institutions in and beyond the forest sector are changing at different paces depending on the larger political framework in the countries. For example, the collapse of the Soviet Union brought major transformations in Central Asia that have had direct and indirect impacts on the forest sector. Institutional capacities have declined, and forest policies, legislation and institutions have not yet been adapted to address new challenges in a decentralized framework. In some areas, conflict-related instability is undermining institutional capacity.

Historically, local community institutions had a key role in resource management, but the advent of government control undermined traditional management systems, often resulting in unregulated resource use (Government of Oman, 2005). While some countries have attempted to broaden participation (Box 26), participatory approaches have not yet taken root in most countries. However, where democratic processes are well established (e.g. Cyprus), forest policies and institutions are responding to society's changing needs, for example, by moving the focus of forest management from wood production to the provision of environmental services and by encouraging participatory approaches.

Private-sector involvement in forest management is limited, largely because most land is publicly owned, and more importantly because productivity and commercial viability are poor. However, in most countries, the private sector is dominant in forest industries and trade in forest products.

BOX 26	Village cooperatives in Turkey

Turkey has about 5 000 agricultural village cooperatives with a total of more than 680 000 members. About 3 200 of these cooperatives are in forest villages. Forest laws have granted special rights and privileges to forest village cooperatives since the 1970s, including priority in undertaking forest harvesting operations and eligibility to take a share of the wood they harvest at reduced rates. More than 2 100 village cooperatives carried out forestry operations in 2000, harvesting about 60 percent of the country's total wood production.

SOURCE: FAO, 2008f.

Science and technology

From 1997 to 2002, average R&D expenditure in the region remained below 0.5 percent of total GDP (FAO, 2007c), significantly lower than the world average even for developing countries. However, the number of Internet users is rising, indicating that access to information is increasing. Most countries in Central Asia benefited from the large science and technology infrastructure base of the Soviet Union, and the scientific capacity of these countries has declined since its collapse. Limited resources, a top-down approach to R&D and the loss of competent scientists through emigration have affected the scientific and technological capabilities of most countries in the region, with the exception of a few such as the Islamic Republic of Iran and Turkey. Overall, forestry has low priority in the region and the sector receives minimal investment. The areas receiving most attention are forest conservation and environmental services.

OVERALL SCENARIO

Three broad patterns of development can be identified in the region, with differing implications for forests and forestry.

A number of non-fossil-fuel-producing, low-income countries will continue to depend on agriculture and animal husbandry as a main source of livelihood (with remittances from citizens employed in fossil-fuel-producing countries also becoming an important source of income). The outlook for forests and woodlands will depend on diversification of the economy – which will depend in turn on political stability, institutional development and investment in human resources. Tourism offers potential for diversification.

Countries that depend on fossil fuels for their growth and prosperity also need to diversify. Several realize the long-term vulnerability of dependence on fossil fuels and are investing in manufacturing and building up human resources. Many of these countries have neglected sectors other than energy, including agriculture and forestry; thus, despite high national income, forestry may face severe financial constraints and forestry institutions may be weak. Improving institutional frameworks is likely to remain a major challenge.

Some countries (both fossil-fuel-producing and non-fossil-fuel-producing) have made substantial progress in diversifying their economies and drawing advantage from globalization through investments in manufacturing, trade, commerce and tourism. Several are emerging as important regional and global financial centres. In these countries, increasing attention is being paid to environmental issues, including urban greening.

OUTLOOK

Forest area

Forest area increased between 1990 and 2005 (Table 18). This trend is likely to continue except in the low-income agriculture-dependent countries. As the importance of agriculture (including animal husbandry) declines and wealthier countries invest in afforestation and urban greening (Box 27), the overall forest area is expected to increase. In countries with low forest cover, rapid urbanization and constraints on agricultural expansion (especially water scarcity), forest area is likely to stabilize. Afforestation efforts, although limited, will help to reverse forest loss. A notable exception to this trend is a continuing decline in forest area in those countries where armed conflict has destabilized forest management.

Rangelands and pasturelands with scattered tree growth account for more than half the region's land area and are the main source of fodder and woodfuel in addition to a number of non-wood tree products. These lands are rapidly becoming degraded in the absence of any management (Box 28).

In addition to difficult-to-quantify but important trees outside forests and in agroforestry systems, the region has about 5 million hectares of planted forests. This is less than 2 percent of the global planted forest area (Table 19). Half of these planted forests are for environmental protection. The annual rate of planting has been rather modest at about 80 000 ha. The decline in the extent of planted forests in Central Asia between 2000 and 2005 was mainly in Kazakhstan and was largely the result of forest fires (FAO, 2006d). Half the region's planted forests are in Turkey, where 75 percent are for production and the rest for protection. The Islamic Republic of Iran and Turkey are the only countries reporting planted forests for production.

TABLE 18

Forest area: extent and change

Subregion	Area (1 000 ha)			Annual change (1 000 ha)		Annual change rate (%)	
	1990	2000	2005	1990–2000	2000–2005	1990–2000	2000–2005
Central Asia	15 880	15 973	16 017	9	9	0.06	0.06
Western Asia	27 296	27 546	27 570	25	5	0.09	0.02
Total Western and Central Asia	**43 176**	**43 519**	**43 588**	**34**	**14**	**0.08**	**0.03**
World	4 077 291	3 988 610	3 952 025	−8 868	−7 317	−0.22	−0.18

NOTE: Data presented are subject to rounding.
SOURCE: FAO, 2006a.

BOX 27	Tree planting in the United Arab Emirates

The United Arab Emirates is an extremely arid and urbanized (over 80 percent) country. The government encourages greening and tree-planting activities, which are increasingly supported by the people.

Urban planting schemes enhance microclimate, mitigate air pollution, beautify roadsides and provide recreational areas. Abu Dhabi, which had only one public park in 1974, now has about 40, covering an area of more than 300 ha.

Outside cities, trees are planted:
- in green belts to combat desertification and sand movement;
- to protect farms, agricultural areas and rangelands;
- to provide natural sanctuaries for the breeding and conservation of gazelles, bush rabbits, birds and other animals.

Ninety percent of treated wastewater is used for irrigation of these planted areas.

SOURCE: FAO, 2005c.

BOX 28	Rangelands in West Asia

Rangelands occupy 52 percent of West Asia's land area. Up to 90 percent of these lands are degraded or vulnerable to desertification. Grazing, a principal cause of land degradation in the subregion, has more than doubled in the past four decades, mainly because of subsidized feeding, provision of water points and mechanization. Sheep density has reached four times the sustainable carrying capacity in some areas. Overgrazing and fuelwood collection have reduced rangeland productivity by 20 percent in Jordan and 70 percent in the Syrian Arab Republic.

Centralized control of rangelands has undermined traditional nomadic herding systems, which managed the land carefully to prevent overuse. Most rangelands in the subregion are free-access resources, lacking clear responsibilities for their protection.

SOURCES: FAO, 2007c; UNEP, 2007.

TABLE 19
Planted forests

Subregion	1990	2000	2005
	(1 000 ha)		
Central Asia	1 274	1 323	1 193
Western Asia	3 022	3 623	3 895
Total Western and Central Asia	4 295	4 946	5 089
World	209 443	246 556	271 346

NOTE: Data presented are subject to rounding.
SOURCE: FAO, 2006b.

Forest management

Except in Cyprus, Lebanon and Yemen, most of the forests in the region are publicly owned. However, political and historical differences among the countries have resulted in considerable differences in how they have been managed and used.

In the Soviet period, most of the forests and woodlands in Central Asia were set aside for environmental protection with a total ban on logging – a policy encouraged by the low forest cover and limited scope for commercial use of forests. Strict enforcement of rules and regulations by the well-organized state forestry administration enabled comprehensive forest protection. However, after independence, a reduction in wood and fuel supplies from the Russian Federation increased the pressure on forests and the ban on logging became ineffective. While most of the forests officially remain protected areas, institutional weaknesses and the rising demand for wood have resulted in increases in illegal logging. Greater investment will be needed if problems such as forest fires are to be prevented from worsening.

In Western Asia too, most forests have been set aside as protected areas. A number of countries that previously depended on forests for wood production have reduced harvesting in order to enhance environmental benefits.

The region's adverse climate and soil conditions and low productivity make forest plantation activities expensive, implying limited private-sector involvement and, thus, a high reliance on public funding. The changing needs of society have influenced the management of planted forests; some originally established for wood production are now managed for amenity values (Box 29).

In most countries in the region, trees grown on farms in various agroforestry systems are a source of income and, more importantly, fulfil protective functions as windbreaks and shelterbelts. Establishment of windbreaks is an integral part of farming practices in most countries. Date-palm cultivation in several Western Asian countries has turned deserts into oases. In the United Arab Emirates, extensive date plantations have improved the landscape while generating substantial income (FAO, 2008f). Fruit trees are also a source of wood.

The high costs of improving policy and institutional arrangements and technical capacities may continue to limit the ability of many countries in the region to implement sustainable forest management. Furthermore, much of the region's forest is located in conflict zones. Conflict-related instability is a major factor undermining sustainable forest management, especially where forests straddle national borders (FAO, 2008g).

Wood products: production, consumption and trade

Because of the unfavourable growing conditions and emphasis on protection, production of wood products is low, and the region depends substantially on imports to meet demand. Imports of wood products increased from about US$5.6 billion in 1995 to US$13.5 billion in 2006 and accounted for more than half of consumption. Afghanistan, Georgia, the Islamic Republic of Iran, Kazakhstan and Turkey account for most of the region's wood production.

Consumption of wood products is forecast to increase across the region with growth in population, urbanization and income. Annual growth in the consumption of sawnwood, wood-based panels, and paper and paperboard is projected to be 2.5, 4.5 and 4.0 percent, respectively, in the next 15 years (Table 20). Growth is expected to be fastest in the Central Asian countries as they recover from the post-1990 economic slump. The region will remain a major wood product importing region because of its limited natural resources and growing demand.

The Islamic Republic of Iran and Turkey, with large domestic markets, inexpensive labour and a stable investment climate, have invested in forest industry development (furniture, paperboard and medium-density fibreboard), largely based on imported raw materials. With the declining profitability of the wood industry in Europe, these industries could expand further. Saudi Arabia and

BOX 29	Changing objectives of forest plantation management in Cyprus

In Cyprus, 94 village plantations on about 1 580 ha were established during the Second World War to supply woodfuel to local communities. By the time the plantations were mature, incomes had increased and commercial fuels had become available and affordable, so the demand for woodfuel had declined considerably. However, the demand for recreation areas had grown. Hence, these plantations were transformed into recreation areas, enhanced by the planting of ornamental trees.

SOURCE: Government of Cyprus, 2005.

TABLE 20
Production and consumption of wood products

Year	Industrial roundwood (million m³)		Sawnwood (million m³)		Wood-based panels (million m³)		Paper and paperboard (million tonnes)	
	Production	Consumption	Production	Consumption	Production	Consumption	Production	Consumption
2000	14	15	6	10	3	6	2	6
2005	17	19	7	13	5	9	3	8
2010	17	21	8	14	6	12	4	10
2020	15	22	10	18	11	18	6	14

SOURCE: FAO, 2008c.

the United Arab Emirates produce paper and paperboard (mainly tissue paper and corrugated carton) almost entirely using imported pulp and locally collected wastepaper. However, the competitiveness of the industry is in question because of the high production costs, especially arising from the high water demand (Mubin, 2004).

Woodfuel

At the aggregate level, woodfuel consumption will continue to decline in the next 15 years (Figure 48). However, consumption trends differ considerably among and sometimes within countries. Turkey, with its diversified economy, has seen a significant reduction in woodfuel use largely because of the availability of commercial fuels, and this trend is likely to continue. However, in low-income countries, commercial fuels are unavailable and the use of woodfuel has increased. For example, woodfuel accounts for almost 85 and 70 percent of household energy needs in Afghanistan and Yemen, respectively (FAO, 2007c). Woodfuel use is also high in some of the Central Asian republics (Tajikistan and Uzbekistan). In these countries, total consumption is projected to rise, which will put additional stress on the low-productivity forests and wooded lands.

In most of the other countries, especially in Western Asia, fuelwood consumption is declining but charcoal use is increasing, particularly in restaurants and homes. In Saudi Arabia, an attempt to conserve the resource by banning charcoal production and encouraging imports was not successful, as people without alternative income opportunities continued to produce charcoal as a source of livelihood.

Non-wood forest products

As in other regions, the pattern of NWFP use consists of many subsistence products and a few commercially important ones, many of which are domesticated and cultivated systematically (FAO, 2006e; FAO, 2007c). Subsistence use of and trade in NWFPs are particularly significant for low-income rural communities. In many countries, NWFPs provide more income than wood production.

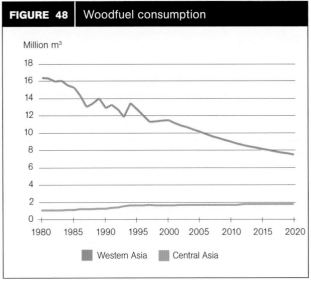

FIGURE 48 | Woodfuel consumption

Million m³

SOURCE: FAO, 2003b.

Commercial products include honey, mushrooms, medicinal plants, pine nuts, walnuts, pistachio nuts, bay leaves, thyme and fodder. In the more diversified economies, commercially important NWFPs have been systematically developed with private-sector involvement. Lebanon's privately owned pine (*Pinus pinea*) plantations are managed primarily for nut production. The production and processing of, and trade in, bay leaves from Turkey have improved largely because of private-sector investments.

No major changes are expected in the pattern of NWFP use. The main challenge will be to improve the production and value addition of less commercialized products, to develop markets and, thus, to enhance income opportunities for low-income households.

Contribution of forestry to income and employment

The gross value added by the forestry sector registered a slight increase from about US$4.9 billion in 1990 to about US$5.3 billion in 2006 (Figure 49). Most of the increase was in the pulp and paper sector, largely because of expanded paperboard production. Employment in the sector has registered an upward trend since 2000 following

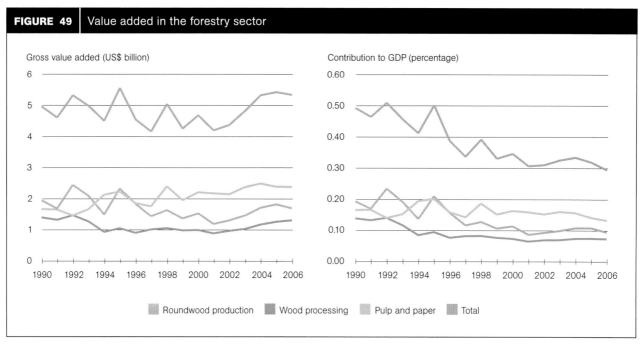

FIGURE 49 | Value added in the forestry sector

Gross value added (US$ billion)

Contribution to GDP (percentage)

Roundwood production ■ Wood processing ■ Pulp and paper ■ Total

NOTE: The changes in value added are the changes in real value (i.e. adjusted for inflation).
SOURCE: FAO, 2008b.

a slight decline, but is essentially stable. However, estimates are imprecise because of incomplete national accounting of value added and employment, especially for the informal sector.

Environmental services of forests

Considering the limited potential of commercial wood production, the provision of environmental services – especially arresting land degradation and desertification, protecting water supplies and improving the urban environment – will remain the principal function of forests and woodlands in Western and Central Asia. Environmental protection and provision of environmental services are largely driven by the public sector through supportive policy measures, with varying levels of participation by civil-society organizations, the private sector and communities.

Five areas in the region have been identified as biodiversity hotspots for their biological richness and threatened ecosystems (Conservation International, 2005). For example, the forests in the Central Asian mountains are the centre of origin of cultivars of apple, pear and pomegranate. To date, biodiversity conservation efforts have focused on designating protected areas, which by 2007 comprised about 114 million hectares or about 10 percent of the region's land area (UN, 2008c).

In low-income agriculture-dependent countries with high biodiversity (e.g. Afghanistan, Kyrgyzstan, Tajikistan and Yemen), conservation may remain difficult because of the pressure on land and other resources and the inability of governments to invest adequately in effective

management of protected areas. Weaknesses in policies and institutions, including fragmented responsibilities, are impediments to protected area management even in some countries with relatively high incomes. Illegal hunting is a major problem in some protected areas.

Desertification and land degradation are problems throughout the region, especially in Western Asia where all countries are in the arid or semi-arid zone and three-quarters of the land is desert or desertified (FAO, 2007c). The causes include extreme climate conditions and human activities, such as expansion of agriculture, intensive grazing, continued removal of vegetation for fuel and fodder and deficient irrigation practices. Forests and trees contribute directly to controlling desertification risks and maintaining suitable conditions for agriculture, rangelands and human livelihoods. However, as trees also consume water, the water balance needs to be taken into account when tree planting is considered; in Israel, it was observed that planting trees on farms may yield more benefits than large-scale afforestation programmes (Malagnoux, Sène and Atzmon, 2007).

Integrated land and water management could prevent human-induced desertification. However, most attention has been focused on remedial measures. Low-income agriculture-dependent countries have relatively poor prospects for dealing with land degradation and desertification. More improvement is envisaged in countries where dependence on land is declining and opportunities for improving policies and institutional framework are greater.

Climate change is expected not only to accentuate desertification but also to affect water supply because of the shrinking of glaciers in the Central Asian mountains.

Water is probably the most critical natural resource in the region. Watershed degradation is a threat to water supply for drinking, irrigation and power generation. Forests and trees have an important role in watershed improvement. The transboundary nature of most of the major watersheds in the region complicates the institutional arrangements for watershed management, including sharing costs and benefits. Water sharing among the countries is a politically sensitive issue and a primary cause of conflicts in the region.

Unspoilt landscape, including mountains and deserts, is attracting an increasing number of domestic and international tourists to the region, creating opportunities as well as challenges. Increasing investment in infrastructure – for example, construction of the New Silk Road – is opening up the hitherto less-visited Central Asian countries. While many of the diversified economies have been able to take advantage of ecotourism (Box 30), several others, especially in Central Asia, have been unable to tap the potential because of limited infrastructure and security issues. Most low-income countries lack the institutional arrangements to ensure that income from ecotourism accrues to the poor.

The main challenge of nature-based tourism is to manage it sustainably. Increased tourism (including domestic tourism) to a small number of prime locations (e.g. the Azir region of Saudi Arabia) challenges existing institutional capacity to do so. Opening the region to nature tourism is also paving the way for illegal trophy hunting, especially where law enforcement capacity is poor (FAO, 2005d).

Most Western and Central Asian countries invest substantially in creating green spaces to improve the quality of life of the burgeoning urban population (FAO, 2005c). In most Central Asian countries, the significant attention given to urban forestry during the Soviet period declined following independence, but it is picking up again, especially in the fossil-fuel-rich countries. As illustrated in Box 27, several GCC countries have embarked on ambitious greening programmes in conjunction with the expansion of urban centres.

Urban green spaces in the region will undoubtedly increase in varying measure depending on the financial and institutional capacity of the countries and the extent to which urbanization is planned. Unplanned urbanization (especially where rural populations are compelled to move to urban centres because of conflict) tends to result in destruction of urban green spaces.

SUMMARY

The outlook for forests and forestry in Western and Central Asia is mixed. Income growth and urbanization suggest a stable or improving forest situation in some countries, but this will elude a number of low-income agriculture-dependent countries. Forest degradation may also persist in some countries that are relatively well off but have weak institutions.

Adverse growing conditions in most of the region's countries limit the prospects for commercial wood production. Rapidly increasing income and high population growth rates suggest that the region will continue to depend on imports to meet the demand for most wood products. Provision of environmental services will remain the main justification for forestry, especially arresting land degradation and desertification, protecting watersheds and improving the urban environment. Institution building, particularly at the local level, is needed to facilitate an integrated approach to resource management.

BOX 30	Ecotourism development in Tajikistan

In the Murgab District in the Eastern Pamir mountains of Tajikistan, where living conditions deteriorated after the collapse of the Soviet Union, the Murgab Ecotourism Association is promoting sustainable ecotourism with a focus on conservation of natural and cultural resources and local income generation. The Murgab Ecotourism Association was established in 2003 by the Agency for Technical Cooperation and Development with assistance from the United Nations Educational, Scientific and Cultural Organization (UNESCO). Since 2005, it has been a legally registered national association. The number of tourists using its services (which include organization of rafting and camelback tours and accommodation in yurts and local homesteads) grew from 25 in 2003 to 601 in 2005. Profits for local tourism operators increased tenfold. Future plans include the establishment of a nationwide ecotourism network, expanded support to the handicraft production chain, collaboration with large-scale commercial tourism providers and the government, and regional links with northern Afghanistan and southern Kyrgyzstan.

SOURCE: ACTED, 2006.

PART 2

Adapting for the future

Demand for goods and services, and consequently the income that owners derive from managing forests, is a primary determinant of investment in forest management. The first chapter in Part 2 looks at long-term changes in demand for wood and selected wood products, from 1965 to 2005 and projecting to 2020 and 2030. The second chapter focuses on the demand for environmental services provided by forests and on the market and non-market mechanisms that have been evolving to help forests meet that demand.

Institutions are at the centre-stage of sustainable resource management and will play an important part in society's adaptation to social, economic and environmental change. The third chapter provides an overview of how the various institutions in the forest sector – public agencies, the private sector, civil-society organizations, the informal sector and international organizations – are responding to the emerging developments outlined in Part 1.

Science and technology have a vital role in enabling society to meet its future needs for products and services. The final chapter of *State of the World's Forests 2009* provides an overview of science and technology developments in forestry that will help the forest sector address the challenges ahead, including recognition of the role of traditional knowledge.

Global demand for wood products

Demand for wood products is one of the main drivers of investment in forest management. Although short-term market changes influence individual decision-making, long-term changes in demand have a greater influence on investments in forestry and forest industry at the aggregate level. This chapter projects some of the long-term changes in the demand for wood products (based on FAO, 2008c).

DRIVERS OF CHANGE

The main factors affecting long-term global demand for wood products include:

- Demographic changes: the world's population is projected to increase from 6.4 billion in 2005 to 7.5 billion in 2020 and 8.2 billion in 2030.
- Continued economic growth: global GDP increased from about US$16 trillion in 1970 to US$47 trillion in 2005 (at 2005 prices and exchange rates) and is projected to grow to almost US$100 trillion by 2030 (Figure 50).
- Regional shifts: developed economies accounted for most of the GDP in the period 1970–2005. The rapid growth of developing economies, especially in Asia, will swing the balance significantly in the next 25 years.
- Environmental policies and regulations: more forests will be excluded from wood production.
- Energy policies: the use of biomass, including wood, is increasingly encouraged.

Other important factors in the wood products outlook include a decline in harvesting from natural forests and the emergence of planted forests as the major source of wood supply (Box 31), and technological developments such as increased plantation productivity through tree improvement, reduced wood requirements owing to expanded recycling, higher recovery, wider use of new composite products and production of cellulosic biofuel (see the chapter "Developments in forest science and technology" in Part 2).

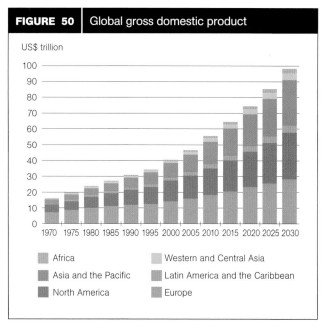

FIGURE 50 | Global gross domestic product

US$ trillion

Africa

Asia and the Pacific

North America

Western and Central Asia

Latin America and the Caribbean

Europe

NOTE: 2005 prices and exchange rates.
SOURCES: FAO, 2008a, 2008c.

OUTLOOK
Sawnwood

Long-term annual growth in sawnwood production and consumption was about 1.1 percent globally in the period 1965–1990, but declined drastically from 1990 to 1995, mostly as a result of reductions in Eastern Europe and the former Soviet Union (Table 21; Figure 51). Sawnwood production and consumption have also declined in Asia and the Pacific since 1995.

Europe and North America account for about two-thirds of global sawnwood production and consumption and are net exporters of sawnwood. Latin America and the Caribbean, the other net exporting region, accounts for almost 10 percent of production, while Asia and the Pacific accounts for slightly more than 15 percent of production and is the world's main net importing region. Production and consumption of sawnwood in Africa and in Western and Central Asia are modest, amounting to less than 5 percent of the global total between them.

| BOX 31 | Outlook for wood production from planted forests |

The world's forest plantation area, as reported to the Global Forest Resources Assessment 2005 (FAO, 2006a), is 140.8 million hectares. The area of planted forests, defined more broadly to include the planted component of semi-natural forests, is estimated to be 271 million hectares (FAO, 2006b).

The outlook for global wood production from planted forests to 2030 was estimated based on a survey of planted forests in 61 countries, representing about 95 percent of the estimated global planted forest area (FAO, 2006b). The outlook was calculated based on predicted changes in planted forest area (mainly through new plantings) as well as opportunities for increased productivity from more efficient management practices, new technology and genetic improvements, following three scenarios:

- Scenario 1: increase in planted forests slowing to half the pace of previous trends (owing to constraints including lack of suitable land and slow growth in demand), with no change in productivity;
- Scenario 2: area changes continuing at the current rate until 2030, without productivity increases;

- Scenario 3: area changes continuing at the current rate until 2030, with an annual productivity increase (for those management schemes where genetic, management or technological improvements are expected).

The model results indicate that the area of planted forests will increase in all scenarios in all regions except Africa, with the highest increase in Asia (figure on the left). Among species groups, the highest increase will be in pine forests.

The total wood volume produced will increase across all scenarios from 2005 to 2030 (figure on the right). The widest variation among scenarios is in Asia and South America, where the higher-productivity Scenario 3 gives a considerable increase in wood production, mainly in eucalyptus and other hardwood species. The differences between Scenarios 1 and 2 are small, as additional planted area in Scenario 2 may not generate wood within the period of the projection.

Actual production could vary significantly from the projections. Often, planted forests are not harvested even on reaching maturity, particularly when they are established without considering access to markets and potential end uses.

Current and projected planted forest area in 61 countries

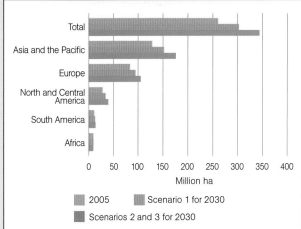

Million ha

Current and projected wood production from planted forests in 61 countries

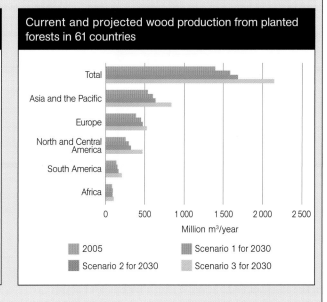

Million m³/year

SOURCE: Carle and Holmgren, 2008.

Timber and the future of tropical forests

From the International Tropical Timber Organization (ITTO)

Large-scale payments for ecosystem services (especially for climate-related services) offer the best prospect for generating funds to secure the tropical forest resource base. However, the main source of income from tropical forests remains timber and wood products. Annual exports of primary and secondary wood products from tropical forests have exceeded US$20 billion in recent years, with further increases foreseen as more countries focus exports on higher-valued secondary wood products.

Much of the raw material already comes from planted forests. The vast areas of degraded forest land in the tropics provide much scope for further increasing planted area, with potential benefits for the wood-processing sector and opportunities for capturing funds from emerging greenhouse gas markets. However, it is important to ensure that payments for ecosystem services do not lead countries to convert natural forest to fast-growing plantations.

Certification and public-purchasing policies are likely to become more important for exporters of tropical wood products in the future, especially as more countries begin to insist on evidence of sustainability, including China (in response to demands from its own export markets). Cellulosic biofuels are likely to provide economic alternatives for tropical countries, but technology transfer from developed countries will be required in order to realize this potential.

The main challenge in the future, as now, will be to add value to tropical forests so that deforestation becomes an economically unattractive option. Despite the potential of new funding mechanisms for tropical forests, it is highly likely that there will be less money available than required.

TABLE 21

Production and consumption of sawnwood

Region	Amount (million m^3)					Average annual change (%)			
	Actual			Projected		Actual		Projected	
	1965	1990	2005	2020	2030	1965–1990	1990–2005	2005–2020	2020–2030
Production									
Africa	3	8	9	11	14	3.7	0.5	1.6	1.9
Asia and the Pacific	64	105	71	83	97	2.0	−2.6	1.1	1.6
Europe	189	192	136	175	201	0.1	−2.2	1.7	1.4
Latin America and the Caribbean	12	27	39	50	60	3.3	2.5	1.7	2.0
North America	88	128	156	191	219	1.5	1.3	1.4	1.4
Western and Central Asia	2	6	7	10	13	4.6	1.5	2.6	2.2
World	**358**	**465**	**417**	**520**	**603**	**1.1**	**−0.7**	**1.5**	**1.5**
Consumption									
Africa	4	10	12	19	26	3.6	1.2	2.8	3.5
Asia and the Pacific	64	112	84	97	113	2.3	−1.9	1.0	1.6
Europe	191	199	121	151	171	0.2	−3.3	1.5	1.2
Latin America and the Caribbean	11	26	32	42	50	3.3	1.5	1.7	1.8
North America	84	117	158	188	211	1.3	2.0	1.2	1.2
Western and Central Asia	3	7	13	18	23	4.0	3.7	2.5	2.2
World	**358**	**471**	**421**	**515**	**594**	**1.1**	**−0.8**	**1.4**	**1.4**

NOTE: Data presented are subject to rounding.
SOURCES: FAO, 2008a, 2008c.

Projections suggest that the distribution of production and consumption among different regions will not change markedly before 2030, but that growth will increase at the global level. Production growth is expected to be highest in the Russian Federation, Eastern Europe and South America. High growth in consumption is expected in Africa and in Asia and the Pacific. These regions, together with Western and Central Asia, will remain dependent on imports to meet their demand. Consumption growth in developed countries is expected to be more moderate because of replacement by engineered (composite) wood products.

Wood-based panels

Although production and consumption of wood-based panels – including plywood, veneer sheets, particleboard and fibreboard – are currently only half those of sawnwood, their higher growth rates will bring them to the levels of sawnwood by 2030 (Table 22; Figure 52). However, future growth in production and consumption will be slightly slower than in the past in most regions, which suggests that the substitution of wood-based panels for sawnwood may be slowing.

Production and consumption are currently evenly balanced among the three main markets (Asia and the Pacific, Europe and North America). Asia and the Pacific will account for a greater proportion of global wood-based panel production and consumption in the future.

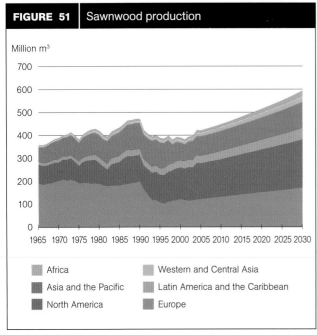

FIGURE 51 Sawnwood production

SOURCES: FAO, 2008a, 2008c.

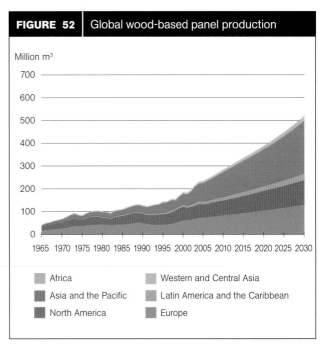

FIGURE 52 Global wood-based panel production

SOURCES: FAO, 2008a, 2008c.

TABLE 22
Production and consumption of wood-based panels

Region	Amount (million m³)					Average annual change (%)			
	Actual			Projected		Actual		Projected	
	1965	1990	2005	2020	2030	1965–1990	1990–2005	2005–2020	2020–2030
Production									
Africa	1	2	3	4	5	4.6	3.8	2.1	2.4
Asia and the Pacific	5	27	81	160	231	6.9	7.5	4.6	3.7
Europe	16	48	73	104	129	4.5	2.8	2.4	2.2
Latin America and the Caribbean	1	4	13	21	29	7.4	7.6	3.3	3.2
North America	19	44	59	88	110	3.4	2.0	2.7	2.2
Western and Central Asia	0	1	5	11	17	6.8	8.9	5.4	4.7
World	**41**	**127**	**234**	**388**	**521**	**4.6**	**4.2**	**3.4**	**3.0**
Consumption									
Africa	0	1	3	4	5	4.8	5.3	1.9	2.4
Asia and the Pacific	4	24	79	161	236	7.4	8.2	4.8	3.9
Europe	16	53	70	99	122	4.9	1.9	2.4	2.1
Latin America and the Caribbean	1	4	9	12	15	7.0	5.7	2.2	2.3
North America	20	43	70	96	115	3.1	3.3	2.1	1.8
Western and Central Asia	0	2	9	18	28	8.1	10.6	4.5	4.5
World	**42**	**128**	**241**	**391**	**521**	**4.6**	**4.3**	**3.3**	**2.9**

NOTE: Data presented are subject to rounding.
SOURCES: FAO, 2008a; FAO, 2008c.

Within the category of wood-based panels, there is an increasing shift from plywood (which accounted for most of the wood-based panel production and consumption in the 1960s) to particleboard and fibreboard. This shift, which has important implications for wood raw-material requirements, began in Europe (where particleboard and fibreboard accounted for 90 percent of the panel market in 2005) and has continued in North America (70 percent). It has only recently started to occur in Asia and the Pacific, where plywood still accounts for more than half of production and consumption, with two main producers (Indonesia and Malaysia) and two main consumers (China and Japan).

Asia and the Pacific, Europe and Latin America and the Caribbean are net exporting regions, while the others are net importers. Europe exports mainly particleboard and fibreboard, while the other two regions export plywood. These trends are expected to continue, with international trade accounting for about 26–27 percent of global production and consumption.

Paper and paperboard

As with panel products, global production of paper and paperboard is also expanding rapidly (although less so than in recent decades), with an annual growth rate of 3.7 percent between 1965 and 1990 and 2.8 percent between 1990 and 2005. Growth rates for consumption have been about the same as those for production (Table 23; Figure 53).

Historically, North America dominated global production and consumption, but because of rapid growth in Asia and the Pacific and Europe, all three major markets now account for a similar share. The rapid growth in Asia and the Pacific is a consequence of the high rate of economic growth in recent decades, first in Japan and a few other industrializing economies and more recently in China and India.

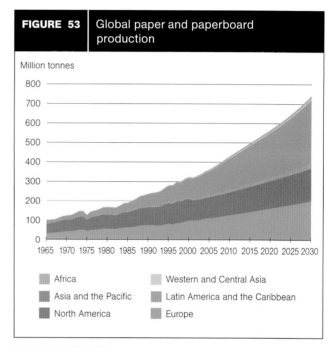

FIGURE 53 Global paper and paperboard production

Million tonnes

Legend:
- Africa
- Asia and the Pacific
- North America
- Western and Central Asia
- Latin America and the Caribbean
- Europe

SOURCES: FAO, 2008a, 2008c.

TABLE 23
Production and consumption of paper and paperboard

Region	Amount (million tonnes)					Average annual change (%)			
	Actual			Projected		Actual		Projected	
	1965	1990	2005	2020	2030	1965–1990	1990–2005	2005–2020	2020–2030
Production									
Africa	1	3	5	9	13	6.4	4.3	3.9	3.7
Asia and the Pacific	13	58	121	227	324	6.3	5.0	4.3	3.6
Europe	33	76	111	164	201	3.4	2.6	2.6	2.1
Latin America and the Caribbean	2	8	14	21	27	5.7	3.6	2.9	2.7
North America	48	91	109	141	169	2.6	1.2	1.8	1.8
Western and Central Asia	0	1	3	6	9	9.2	5.9	4.2	3.5
World	**96**	**238**	**363**	**568**	**743**	**3.7**	**2.8**	**3.0**	**2.7**
Consumption									
Africa	1	4	7	14	21	5.1	4.2	4.6	4.4
Asia and the Pacific	13	63	128	234	329	6.3	4.9	4.1	3.5
Europe	32	73	101	147	180	3.3	2.2	2.6	2.0
Latin America and the Caribbean	3	9	16	24	31	4.7	3.9	2.9	2.6
North America	46	87	106	138	165	2.6	1.3	1.8	1.8
Western and Central Asia	0	3	8	14	20	7.5	7.5	4.0	3.4
World	**96**	**237**	**365**	**571**	**747**	**3.7**	**2.9**	**3.0**	**2.7**

NOTE: Data presented are subject to rounding.
SOURCES: FAO, 2008a, 2008c.

In Europe, production growth has been driven partly by the expansion of exports; Europe is the largest exporter of paper products. On the supply side, European production has also benefited from high growth in wastepaper recovery.

The differences in past and future growth also reflect the structure of the paper and paperboard markets and industry in the three main regions:

- Currently, global newsprint production is divided roughly equally among Asia and the Pacific, Europe and North America, but growth is slowing because of the rapid spread of electronic media.
- Asia and the Pacific and Europe produce far more printing and writing paper than North America.
- Production of other paper and paperboard is highest in Asia and the Pacific.

Paper and paperboard is one of the most globalized commodity groups, with a high share of production exported and a high share of consumption imported. International trade expanded significantly in the 1990s, especially in Europe, and the globalization of paper and paperboard markets will continue in the future.

Industrial roundwood

Industrial roundwood demand is derived from growth in demand for end products – sawnwood, wood-based panels and paper and paperboard. Wood requirements for these products vary depending on the technology employed and the potential to use wood and fibre waste. Growth in sawnwood production requires more industrial roundwood, whereas a shift to reconstituted panel production (particleboard and fibreboard) increases the potential to use wood residues and fibre waste, reducing industrial roundwood requirements. Recycling policies have led to increased use of recovered paper and reduced pulpwood demand.

Increased use of wood residues and recycled materials will reduce the share of industrial roundwood in total wood and fibre use from almost 70 percent in 2005 to about 50 percent in 2030.

The total derived demand in wood raw-material equivalent (WRME) is higher than the consumption of industrial roundwood. In 2005, global derived demand amounted to about 2.5 billion cubic metres WRME, of which 1.7 billion cubic metres was industrial roundwood. Approximately 0.5 billion cubic metres WRME came from recovered paper and the remainder from wood-processing residues, recovered wood products and other sources.

Global production of industrial roundwood is expected to increase by slightly more than 40 percent up to 2030 (Table 24; Figure 54). This is considerably less than the projected rise in total wood and fibre demand (which is expected to almost double) because the highest rates of production growth are expected in the paper and paperboard sector and a higher proportion of paper consumption will be recycled in the future.

TABLE 24
Production and consumption of industrial roundwood

Region	Amount (million m³)					Average annual change (%)			
	Actual			Projected		Actual		Projected	
	1965	1990	2005	2020	2030	1965–1990	1990–2005	2005–2020	2020–2030
Production									
Africa	31	55	72	93	114	2.4	1.8	1.8	2.0
Asia and the Pacific	155	282	273	439	500	2.4	−0.2	3.2	1.3
Europe	505	640	513	707	834	0.9	−1.5	2.2	1.7
Latin America and the Caribbean	34	114	168	184	192	5.0	2.6	0.6	0.4
North America	394	591	625	728	806	1.6	0.4	1.0	1.0
Western and Central Asia	10	9	17	15	11	−0.6	4.5	−0.8	−3.0
World	**1 128**	**1 690**	**1 668**	**2 166**	**2 457**	**1.6**	**−0.1**	**1.8**	**1.3**
Consumption									
Africa	25	51	68	88	109	2.9	1.9	1.8	2.1
Asia and the Pacific	162	315	316	498	563	2.7	0.0	3.1	1.2
Europe	519	650	494	647	749	0.9	−1.8	1.8	1.5
Latin America and the Caribbean	33	111	166	181	189	4.9	2.7	0.6	0.4
North America	389	570	620	728	808	1.5	0.6	1.1	1.0
Western and Central Asia	10	10	19	22	19	−0.2	4.4	1.1	−1.3
World	**1 138**	**1 707**	**1 682**	**2 165**	**2 436**	**1.6**	**−0.1**	**1.7**	**1.2**

NOTE: Data presented are subject to rounding.
SOURCES: FAO, 2008a; 2008c.

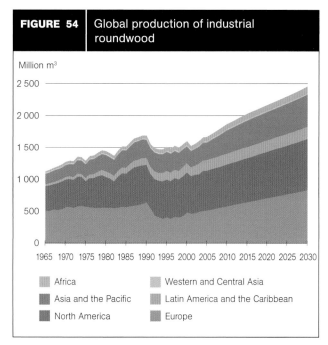

FIGURE 54 | Global production of industrial roundwood

Million m³

Africa	Western and Central Asia
Asia and the Pacific	Latin America and the Caribbean
North America	Europe

SOURCES: FAO, 2008a, 2008c.

Most of the growth will occur in the three main regional markets. The greatest production expansion will be in Europe (more than 300 million cubic metres), mostly because of increases in the Russian Federation. Production in Asia and the Pacific and North America will also expand, largely because of increased production from planted forests.

Asia and the Pacific will have a high deficit between production and consumption, increasing from about 43 million cubic metres in 2005 to 63 million cubic metres in 2030. Thus, the region will depend on potential surplus countries, especially the Russian Federation and possibly some countries in Latin America and the Caribbean.

In the 1990s, Europe, which had been a net importer of industrial roundwood, became a net exporter, largely because of exports from the Russian Federation. The opposite trend was observed in Asia and the Pacific. This situation is likely to continue in the future, although it could be influenced by recent changes in the Russian Federation's forest policies (see Box 10 on page 26).

Wood energy

Roundwood used in energy production is comparable in quantity with industrial roundwood. Energy production using wood includes traditional heating and cooking with fuelwood and charcoal, heat and power production in the forest industry (usually using processing wastes such as black liquor from pulp production) for own use or sale to others, and heat and power generation in specifically designed power facilities.

Statistics on energy production from wood are difficult to obtain because of this diversity of uses and the high

share of informal production. Furthermore, the two main agencies that collect these statistics – FAO and the International Energy Agency (IEA) – present different figures because of different definitions and primary data sources. IEA presents biomass energy production figures that include other types of biomass besides wood (i.e. agricultural residues and dung). Its statistics also include heat and power generation in the forest industry and by commercial energy producers, which are not fully captured in FAO statistics.

Trends and projections for biomass energy production estimated from a combination of these two data sources reveal an increase in global production from about 530 million tonnes oil equivalent (MTOE) in 1970 to about 720 MTOE in 2005, projected to reach 1 075 MTOE in 2030 (Table 25; Figure 55).

Interpolation suggests that wood used for bioenergy production increased from about 2 billion cubic metres in 1970 to 2.6 billion cubic metres in 2005. This suggests that up to 3.8 billion cubic metres of wood could be required by 2030. However, some of the future demand may be satisfied by biomass produced from agricultural residues and energy crops (including short-rotation coppice and grasses).

Until 2005, global biomass energy production increased relatively slowly, at less than 1 percent per year. Most of the increase in production occurred in developing countries, where wood continues to be a major source of energy. The exception is Asia and the Pacific, where growth has declined considerably because of switching to other preferred types of energy as a result of increasing income.

The projections reflect a future marked increase in the use of biomass for energy production in Europe and, to a lesser extent, North America as renewable energy policies and targets take effect. Europe's per capita biomass energy use is projected to triple by 2020 in response to renewable energy targets, although some production will also come from energy crops and agricultural residues. Most developed countries have set renewable energy targets for 2020; hence, rapid growth in production is expected until that time, followed by a slower rate of growth.

Furthermore, future large-scale commercial production of cellulosic biofuel could increase the demand for wood drastically, beyond that shown in the projections.

The projections for biomass energy production in developing countries also have interesting features:

• In Africa, the growth in biomass energy production will continue, but will slow significantly. With the region's relatively small processing sector and few renewable energy targets, most of its bioenergy production will continue to be from traditional woodfuel (fuelwood and charcoal). Following the trend in other regions (e.g. Asia and the Pacific), this

TABLE 25
Production of bioenergy

Region	Amount (MTOE)[1]					Average annual change (%)			
	Actual			Projected		Actual		Projected	
	1970	1990	2005	2020	2030	1970–1990	1990–2005	2005–2020	2020–2030
Africa	87	131	177	219	240	2.1	2.0	1.4	0.9
Asia and the Pacific	259	279	278	302	300	0.4	0.0	0.6	−0.1
Europe	60	70	89	272	291	0.7	1.6	7.7	0.7
Latin America and the Caribbean	70	88	105	123	133	1.1	1.2	1.1	0.8
North America	45	64	65	86	101	1.8	0.1	2.0	1.6
Western and Central Asia	11	7	6	8	10	−2.7	−1.0	2.4	1.9
World	**532**	**638**	**719**	**1 010**	**1 075**	**0.9**	**0.8**	**2.3**	**0.6**

[1] MTOE = million tonnes oil equivalent.
NOTE: Data presented are subject to rounding.
SOURCES: FAO, 2008a, 2008c.

growth is expected to decline as incomes rise and more people switch to other types of energy.
- In Asia and the Pacific, traditional woodfuel production is expected to decline, but this will be outweighed by increased production of bioenergy in the forest industry and, in a few cases (e.g. China), commercial bioenergy production in response to renewable energy targets.
- In Latin America and the Caribbean, biomass energy production is projected to increase in all dimensions, with a rise in traditional woodfuel production in the poorer countries of the region and increased bioenergy production by the forest industry and others in the more advanced economies.

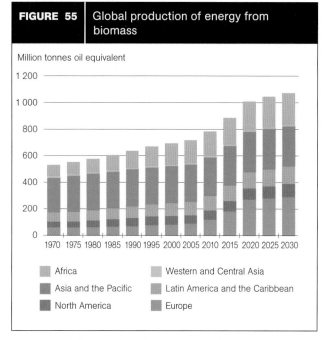

FIGURE 55 | Global production of energy from biomass

Million tonnes oil equivalent

Africa
Asia and the Pacific
North America
Western and Central Asia
Latin America and the Caribbean
Europe

NOTES: 1 tonne of oil equivalent is equal to approximately 4 m³ of wood.
Figures include the use of black liquor, agricultural residues and dung in addition to wood.
SOURCES: FAO, 2008a, 2008c.

SUMMARY

The production and consumption of wood products and wood energy are expected to increase, largely following historical trends. One shift will be the higher growth in the production and consumption of wood products in Asia and the Pacific, mainly stemming from the rapid growth in demand from emerging economies such as China and India. The most dramatic change will be the rapid increase in the use of wood as a source of energy, particularly in Europe as a result of policies promoting greater use of renewable energy.

The Asia and the Pacific region is becoming the major producer and consumer of wood-based panels and paper and paperboard (although per capita consumption will remain higher in Europe and North America). The region's industrial roundwood production will be far short of consumption, increasing dependence on imports unless substantial efforts are made to boost wood production. However, it will be difficult to expand wood production in Asia and the Pacific given the high population density and competing land uses.

Changes in the use of wood for energy and particularly the potential for large-scale commercial production of

cellulosic biofuel will have unprecedented impacts on the forest sector. Increasing transport costs could also influence these projections. Most of the growth in global forest products value chains has been founded on the drastic decline in transport costs in the past two decades. These factors and others, including changes in exchange rates, will influence the competitiveness of the forest sector and affect the production and consumption of most forest products.

Furthermore, the industrial roundwood that is used is increasingly likely to come from planted forests, as growth in production from planted forests is expected to keep up with demand growth for industrial roundwood. This presents interesting opportunities and challenges for management of the remaining forest estate.

Gross value added in forestry

In 2006, the forest industry contributed approximately US$468 billion or 1 percent of the global gross value added. Although this represents an increase in the absolute value of about US$44 billion since 1990, the share of the forestry sector has declined continuously because of the much faster growth of other sectors (see figure). Between 1990 and 2006, value addition increased significantly in the wood-processing subsector, rose marginally in roundwood production and remained stable in pulp and paper, which accounted for nearly 43 percent of the forestry sector's value added in 2006.

Asia and the Pacific registered the most significant increase in gross value added, a large part of it in the pulp and paper subsector (see table). Its share of roundwood production was relatively stable. Growth in Latin America and the Caribbean was also strong, mostly as a result of expansion in roundwood production. Roundwood production also accounted for the increase in Africa. The increase in North America was mainly in the wood-processing sector, while the pulp and paper sector remained stable. Forestry's value added fell only in Europe, mainly owing to a decline in the pulp and paper subsector. Value added in Western and Central Asia remained stable.

These trends are likely to continue in the next few years, especially as investments in wood production and processing increase in Asia and the Pacific and in Latin America and the Caribbean.

Gross value added

Region	Roundwood production (US$ billion)		Wood processing (US$ billion)		Pulp and paper (US$ billion)		Total (US$ billion)		Contribution to GDP (%)	
	1990	2006	1990	2006	1990	2006	1990	2006	1990	2006
Africa	6	9	2	2	3	3	11	14	1.7	1.3
Asia and the Pacific	29	33	21	30	40	56	90	119	1.4	1.0
Europe	27	25	57	57	74	60	159	142	1.4	1.0
Latin America and the Caribbean	13	21	6	7	11	12	30	40	2.0	1.9
North America	21	27	35	53	73	67	129	147	1.4	1.0
Western and Central Asia	2	2	1	1	2	2	5	5	0.5	0.3
World	**98**	**118**	**123**	**150**	**202**	**201**	**424**	**468**	**1.4**	**1.0**

NOTE: Data presented are subject to rounding.

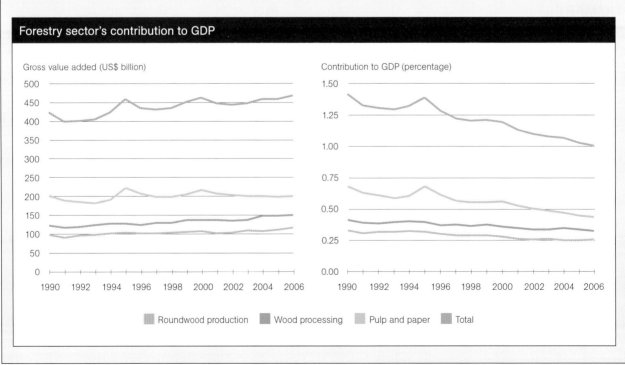

Forestry sector's contribution to GDP

Gross value added (US$ billion)

Contribution to GDP (percentage)

Roundwood production   Wood processing   Pulp and paper   Total

Meeting the demand for environmental services of forests

As the demand for food, fibre and fuel has increased, so has the demand for clean air and water, unspoilt landscapes and other environmental services provided by forests. Where forests are converted to other land uses, the services they supply are diminished. Maintaining such services poses challenges, especially where trade-offs between the production of goods and the provision of services must be addressed.

Publicly owned forests have been a major source of environmental services, provided mainly through regulatory, non-market approaches such as protected areas. With non-state actors playing an increasing role in resource management, a need for incentives for the provision of environmental services has become evident. This chapter discusses the outlook and challenges in the provision of environmental services from forests.

REGULATORY APPROACHES
Protected areas

Establishment of protected areas has been an important and widely adopted regulatory approach to protecting the environment. A main objective is to restrict or prohibit activities that undermine the supply of environmental services. Protected areas are grouped into different categories depending on the degree of protection afforded.

The extent of terrestrial protected areas (including but not only forest protected areas) has registered significant growth in the past three decades, although it seems to have been levelling off since 2000 (Figure 56). The total extent of protected areas is about 1.9 billion hectares, or about 14.5 percent of global land area. This represents an increase of 35 percent since 1990 (UN, 2008c). The area protected varies considerably among the regions. The outlook for protected area management depends on both the scope for increasing the extent of protected areas and the effectiveness of their management.

About 13.5 percent of the world's forests are in some category of protected area (Schmitt *et al.*, 2008). With the exception of some of the large forested regions where population densities are low – the Amazon Basin, the Congo Basin and the boreal forests of Canada and the Russian Federation – the scope for further expansion of protected areas is probably limited.

Effective management of protected areas poses enormous challenges. Much depends on the willingness and ability of society to meet the direct and indirect costs of their management.

In densely populated countries, protected areas are vulnerable to degradation caused by illegal logging, woodfuel collection, grazing and poaching. The ineffectiveness of excluding people has led to a shift in management approach, favouring people's participation in protected area management, including income-sharing arrangements with local communities. The success of such approaches depends on establishing appropriate trade-offs between conflicting objectives. This requires a robust institutional framework and good mediation skills to negotiate a lasting compromise.

Protected areas are often the last frontier for large-scale developments, especially involving mining, oil drilling, infrastructure and large-scale agriculture. Low-income

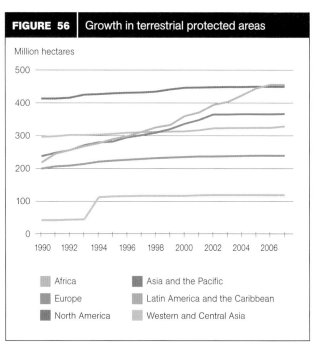

FIGURE 56 | Growth in terrestrial protected areas

Million hectares

- Africa
- Europe
- North America
- Asia and the Pacific
- Latin America and the Caribbean
- Western and Central Asia

SOURCE: UN, 2008c.

countries dependent on land and other natural resources for their development often find it extremely difficult to resist such options.

Sustainable forest management

As less than one-seventh of the world's forests are set aside as protected areas, most forest environmental services are provided in conjunction with the production of wood and other products. Production can be compatible with provision of environmental services, but only up to a certain level. Thus, considerable attention has been devoted to developing wood production systems that minimize environmental damage and support continued provision of services. Implementation of sustainable forest management – which addresses the economic, social and environmental functions of forests – is an important approach to ensuring a balance between the objectives of production and conservation. Maintaining critical ecosystem functions is a key pillar of sustainable forest management. "Close-to-nature silviculture" and the "ecosystem approach" are essentially variants of sustainable forest management, giving greater emphasis to environmental services.

While the concept of sustainable forest management is accepted as the framework for managing forests in most countries, its implementation differs considerably among them. Barriers to its adoption are relatively few where institutions are well developed and society is able to meet the higher costs, as is the case in many developed countries. However, in low-income situations, sustainable forest management faces far more constraints, reflecting limited ability and willingness to pay for the additional costs involved in adhering to the social and environmental criteria. Consequently, in the tropics, the proportion of forests that are sustainably managed remains very low (ITTO, 2006).

Green public procurement

Public procurement policies that aim to ensure that wood products purchased have been produced legally have the potential to promote sustainable forest management and environmental protection. For example, Japan, New Zealand and several countries in Europe have

BOX 32	Green building in the United States of America

"Green building" is construction that conserves raw materials and energy and reduces environmental impacts. It includes consideration of future water use and energy demands, ecological site selection and the procurement of sustainably produced materials. In the United States of America, many public agencies and schools have adopted green building standards. Leadership in Energy and Environmental Design is a green building rating system developed in 1994 by the United States Green Building Council (a member of the World Green Building Council, which has members in more than ten countries). It is a national third-party certification programme for the design, construction and operation of high-performance green buildings. Green building legislation, policies and incentives are in place in 55 cities, 11 counties and 22 states.

While green building provides healthier work environments at both the environmental and human levels, the high costs involved are frequently a disincentive. However, the initial costs are often mitigated over time by gains in overall efficiency.

SOURCE: USGBC, 2008.

operational timber procurement policies, and many regional and local governments have established restrictive rules for their procurement contracts (UNECE and FAO, 2006a). An increasing number of public- and private-sector players are also adopting green building and procurement policies (Metafore, 2007) (Box 32).

MARKET MECHANISMS: THE DEMAND SIDE
Certification for green products

A major condition for the adoption of sustainable forest management is a demand for products that are produced sustainably and consumer willingness to pay for the higher costs entailed. Certification represents a shift from regulatory approaches to market incentives to promote sustainable forest management. By promoting the positive attributes of forest products from sustainably managed forests, certification focuses on the demand side of environmental conservation.

In 2008, more than 300 million hectares, or almost 8 percent of the world's forests, were certified by independent third parties, a significant increase since third-party certification was introduced in 1993 (Figure 57). The two major certification systems are those of the Forest Stewardship Council (FSC) and the Programme for the Endorsement of Forest Certification Schemes (PEFC). In addition, many countries have national certification systems, often affiliated with PEFC (UNECE and FAO, 2006b; ITTO, 2008).

In 2006, certified forests supplied about 24 percent of the global industrial roundwood market (UNECE and FAO, 2006b). FSC (2008) estimates annual sales of FSC-labelled products at US$20 billion. PEFC estimates that 45 percent of the world's roundwood production will come from certified forests by 2017 (Clark, 2007). In addition to wood, other products are increasingly being certified, including woodfuel and NWFPs (UNECE and FAO, 2007).

Both major certification systems now allow non-certified wood to be sold together with certified wood under a "mixed sources" label, provided it meets certain basic requirements of acceptable forest management (World Resources Institute, 2007).

The following are key issues and trends in certification:
- Although certification started with the objective of encouraging sustainable forest management in the tropics, only 10 percent of the certified forest area in 2008 was in the tropics. The rest was in Europe and North America, reflecting economic and institutional advantages in adopting certification in developed countries.
- Certification provides access to markets where consumers prefer green products, but no price premium to cover the costs of certification. For many producers, access to the green market is insufficient incentive for seeking certification, especially when there is demand for comparable uncertified products produced at a lower cost.
- Major expansion in certification will depend on the response of consumers in rapidly growing markets (especially China and India). While the desire for market access may encourage the growth of certification, the main constraints could be on the supply side, especially the investments required to reach the minimum threshold level of management allowing certification.

MARKET MECHANISMS: THE SUPPLY SIDE

Encouraging the supply of environmental services through appropriate payments to forest owners has received considerable attention as a means of supporting forest conservation. While such payments have long existed for recreational services (for example, through entry fees to recreational sites), they are being adopted for other services such as watershed protection, biodiversity conservation and carbon sequestration (Box 33). The idea is to place environmental services on a par with other marketed products, correcting the bias against their supply.

Payments for environmental services (PES) have been developed mainly for watershed services, carbon sequestration and to some extent biodiversity conservation. The growth of ecotourism has also facilitated the development of markets for scenic and nature values, especially through access fees and permits.

Watershed protection

Watershed protection is one of the most important environmental services involving forests and has received considerable attention for payment schemes. These schemes involve payments to upstream land users for improving water quality and quantity through appropriate land-use practices. Such arrangements tend to be most effective in small watersheds, where service providers and beneficiaries are able to interact and the information flow is relatively smooth. At larger scales, more complex arrangements become necessary. In most cases, the payments are from utility companies to land users.

As water is indispensable and tangible, users are generally willing to pay for improving the quality, quantity and regularity of its supply. Moreover, it is easy geographically to identify the providers and beneficiaries of the service.

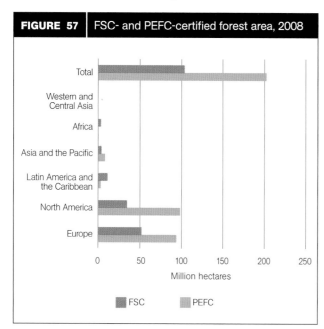

| FIGURE 57 | FSC- and PEFC-certified forest area, 2008 |

SOURCES: FSC, 2008; PEFC, 2008.

BOX 33	Key lessons on developing payment for environmental services schemes

- An operational PES scheme can take years to develop. The crucial step is finding willing buyers.
- Most voluntary, private-driven PES schemes have been small, have high transaction costs and deliver modest rural incomes and modest conservation gains.
- Government-driven PES schemes have tended to be larger and deployed faster, and they have resulted in improved forest practices in some instances.
- Regulation-driven PES with private buyers (e.g. markets for carbon credits) have generated high expectations that have yet to be fully realized.

- PES schemes require supportive legal and institutional frameworks, clear property rights and assistance to small farmers and rural communities.
- National governments remain the most important source of funding for PES programmes, with the international community acting as a catalyst.
- Ecosystem service payments may be insufficient to provide incentives for forest conservation where there are high opportunity costs for land.

SOURCE: FAO, 2007d.

Views from CPF partners

Forests and synergies among multilateral environmental agreements
From the Global Mechanism of the United Nations Convention to Combat Desertification (UNCCD)

The UNCCD promotes synergies offered by forests among multilateral environmental agreements. Sustainable forest management, sustainable land management and climate change adaptation strategies are interrelated; solutions for forest degradation and deforestation overlap with those for land degradation. The Global Mechanism uses national policy processes for coordination and reconciliation, with the aim of increasing investments and financial flows in forests and agriculture. It supports efforts to increase resource allocations in national budgets, to take full advantage of innovative financial mechanisms and to obtain "vertical funds" focused on specific themes.

From a financing perspective, the potential for increased financial flows to address land degradation and degraded forests in the future climate regime is interesting but demands careful preparation. A responsible pro-poor policy framework would provide equitable compensation to smallholder farmers that offer environmental services to the country and climate change resilience to the world. Subsistence farmers in fragile ecosystems could become key players in the international market.

Although forests in arid and semi-arid lands have comparatively low carbon values, they are being degraded at relatively high rates in some regions and, therefore, are targets of national and international schemes. In addition, low-carbon-density forest lands may act as buffer areas between agricultural lands and more dense forests. Their protection is particularly important in preventing encroachment, conversion, further land degradation and eventual desertification. ▓

Views from CPF partners

Valuing ecosystem services
From the United Nations Environment Programme (UNEP)

Climate change poses a major challenge to forests. Its impacts on the supportive and regulating processes of forests and on how people use forest resources are difficult to predict. The best response to the uncertainty climate change presents is to maintain or increase the functioning and resilience of all forests as a matter of urgency. This challenge provides opportunities for forest stakeholders at the national and international levels to increase cooperation.

UNEP promotes an ecosystem approach that considers lessons learned from the past and seeks preparedness for challenges such as climate change. The services that forests provide need to be part of development strategies and incorporated into financial decision-making. Climate regulation is just one of the services for which a monetary value urgently needs to be established. Others include hydrological regulation, protection from natural hazards, nutrient cycling, energy, waste treatment and freshwater provisioning.

As population growth persists and the decline of forest ecosystem services continues, UNEP will promote equitable distribution of ecosystem services across socio-economic groups as an important measure for increasing human well-being and for mitigating conflicts and disasters. ▓

Nevertheless, developing a system of payments for watershed services entails a number of challenges, such as:

- lack of clarity about the hydrological processes involved and, in particular, the impact of different land uses on the quantity, quality and regularity of water flow;
- public opposition related to privatization, perceptions that access to water is a fundamental right and concerns about the potential for increased inequities (i.e. that payment systems might impede poor people's access to water);
- high transaction costs for the development of PES, especially for large watersheds with many providers and users of watershed services.

Consequently, market mechanisms for the provision of watershed services are still in the early stages of development. Most of the existing arrangements are either between small groups of users and providers that can interact efficiently, or established by large electricity or water utilities that can levy the necessary charges and channel the funds to those undertaking watershed conservation.

Carbon markets and forestry

Payment for carbon sequestration to mitigate climate change is one of the fastest-growing environmental markets. Under the Kyoto Protocol, three flexible mechanisms were created: the Clean Development Mechanism (CDM), joint implementation and emission trading. Under the CDM, Annex I (industrialized) countries may offset a certain part of their emissions through investment in carbon sequestration or substitution projects in non-Annex I (developing) countries and thus acquire tradable certified emission reductions. Under joint implementation, Annex I countries may jointly execute carbon sequestration or substitution projects. Emission trading permits the marketing of certified emission reductions.

Carbon markets comprise the compliance market (which follows stringent rules under the Kyoto Protocol) and the voluntary market. In 2007, the total carbon market (including all voluntary and compliance markets) amounted to US$64 billion, more than double the 2006 total (Hamilton *et al.*, 2008). The voluntary carbon market, where a sizeable share of carbon credits comes from forest activities, also doubled in terms of emissions traded (65 million tonnes of carbon dioxide equivalent in 2007), and tripled in terms of value (US$331 million) (Box 34).

While the appeal of afforestation and reforestation as a climate change mitigation strategy is considerable, forest-based carbon offset projects face several challenges, including setting baselines, permanence, leakage and monitoring constraints. The problems are particularly severe in countries with high deforestation rates, which usually also have major policy and institutional constraints. These issues have hindered a more prominent role for forests in climate change mitigation under the CDM (one reforestation project out of 1 133 registered projects as of August 2008).

Following the thirteenth session of the Conference of the Parties to UNFCCC in Bali, Indonesia, in 2007, many high hopes were generated on the inclusion of REDD in the post-Kyoto climate change mitigation efforts. The economic

Reducing emissions from deforestation and forest degradation

From the United Nations Framework Convention on Climate Change (UNFCCC)

Reduction in emissions from deforestation and forest degradation is generally recognized as a relatively low-cost greenhouse gas mitigation option. About 65 percent of the total mitigation potential of forest-related activities is located in the tropics, and about 50 percent of the total could be achieved by reducing emissions from deforestation (IPCC, 2007) – which would also provide other benefits and complement the aims and objectives of other multilateral environmental agreements while addressing some of the needs of local and indigenous communities.

At the Climate Change Conference in Bali, Indonesia, in December 2007, countries adopted a decision on reducing emissions from deforestation in developing countries. Governments are encouraged to seek to overcome the barriers to implementation (lack of effective institutional frameworks, adequate and sustained financing, access to necessary technology and/or appropriate policies and positive incentives) through capacity building, provision of technical assistance, demonstration activities and mobilization of resources.

Several governments have already announced their willingness to support such activities, to provide funds and to address outstanding methodological issues (related to assessment of changes in forest cover and associated forest carbon stocks and greenhouse gas emissions, reference emission levels, estimation of emissions from forest degradation, implications of national and subnational approaches, etc.). Several organizations have also launched initiatives to assist developing countries in these efforts. Opportunities for collaboration should be explored to ensure that efforts are complementary and to maximize the benefits for all countries involved. ▨

and scientific rationale for REDD has been well articulated in that the forest sector (mainly deforestation) accounts for more than 17 percent of greenhouse gas emissions and that addressing deforestation and degradation would be a more cost-effective mitigation option than bringing about changes in energy use. However, providing incentives to desist from deforestation involves complex policy, institutional and ethical issues (Martin, 2008).

Biodiversity conservation

Biodiversity conservation has largely been in the public domain, primarily through establishment and management of protected areas. However, as public funding becomes insufficient to support biodiversity conservation, many

countries have made efforts to identify alternative ways to finance it, including through systems of payment for the services provided. Such systems are compatible with objectives of increased community participation in biodiversity conservation. Examples include private protected areas, which depend on visitor fees as the main source of income.

Payment systems for conservation are diverse (Jenkins, Scherr and Inbar, 2004), including:

- outright purchase of high-value habitat;
- payment for access to potentially commercial species or habitat;
- payment in support of management that conserves biodiversity;

| **BOX 34** | Forests and voluntary carbon markets |

Voluntary carbon markets, or the exchange of offsets by entities not subject to greenhouse gas emission caps, have two components:

- the structured and monitored cap-and-trade system of the Chicago Climate Exchange (CCX);
- the more disaggregated over-the-counter (OTC) system, which is not driven by an emissions cap and does not typically trade on a formal exchange.

In 2007, 42.1 million tonnes of carbon dioxide equivalent (CO_2e) were transacted on the OTC market and 22.9 million tonnes on the CCX, representing a tripling of

transactions for the OTC market and more than a doubling for the CCX since 2006.

Within the larger OTC voluntary market, forestry projects (which include afforestation and reforestation of both planted and natural forests and avoiding deforestation efforts) accounted for 18 percent of transactions in 2007, down from 36 percent in 2006. Projects for avoiding deforestation increased from 3 percent of the volume in 2006 to 5 percent in 2007. Forestry projects and particularly those involving afforestation or reforestation remained among the highest-priced project types in 2006 and 2007, with weighted average prices of US$6.8–8.2 per tonne of CO_2e.

SOURCES: Gorte and Ramseur, 2008; Hamilton *et al.*, 2008.

Views from CPF partners

Primary forests, planted forests and biodiversity objectives

From the Convention on Biological Diversity (CBD)

The year 2010 will be celebrated the world over as the International Year of Biodiversity. This occasion should be used as a starting point for a more sustainable relationship with our forests.

Forests are home to two-thirds of all terrestrial species. If we are to achieve the 2010 target to reduce the loss of biodiversity significantly, all governments and relevant organizations must redouble their efforts to halt deforestation and to manage forests sustainably. For example, market failures that stand in the way of appreciating the real value of forests need to be addressed. Biodiversity and the numerous ecosystem services that forests provide must be properly accounted for, and they must be marketed. Forest governance must be improved and the management of forests must become a matter of societal choice. In addition, information about the

importance and value of forests must reach key decision-makers. The CBD programme of work on forest biodiversity (which was reviewed by the ninth meeting of the Conference of the Parties in Bonn, Germany, in May 2008) addresses all of these issues.

In a context of rising demand for wood products, planted forests will meet a greater part of timber needs in the future. Hence, it is important to ensure that planted forests increasingly fulfil biodiversity objectives, for example by forming ecological corridors between protected areas. New methods and technologies will make it possible to establish planted forests exclusively on degraded lands, without damage to primary forests. Primary forests will serve mostly as reservoirs for biodiversity and as storage space for carbon.

- tradable development rights;
- support for enterprises that adhere to conservation principles in their business practices.

Each of these requires a specific policy and institutional framework.

The market for biodiversity conservation is still nascent. Most of the purchases of high-value habitats (often under debt-for-nature swaps) are by international agencies including non-governmental organizations (NGOs) and foundations. Conservation easements, under which private landowners surrender certain development rights to provide environmental benefits in perpetuity in return for compensation (including tax exemptions), are widely adopted in the United States of America (TNC, 2004).

Other compensation arrangements

Some countries, when unable to avoid the development of forests or other habitats, compensate for the loss by supporting conservation in other locations. Such arrangements involve transfer payments that are not necessarily linked to the quantity or quality of the service delivered and are not true markets for the provision of environmental services in the conventional sense. A typical example is wetland mitigation banking in the United States of America, in which unavoidable impacts on aquatic resources are compensated by establishing, enhancing or conserving another aquatic resource area (US EPA, 2008).

Another example is the compensatory afforestation programme in India under which any diversion of public forests for non-forestry purposes is compensated through afforestation in degraded or non-forest land. Funds received as compensation are used to improve forest management, including afforestation, assisted natural regeneration, management and protection of forests, and watershed management. A government authority has been created specifically to administer this programme (SME Toolkit India, 2008).

SOCIO-ECONOMIC ASPECTS

Support for provision of environmental services and the appropriateness of regulatory and other measures need to be considered in the larger socio-economic context. Countries and societies with higher incomes tend to be more willing to pay for environmental services. Low-income countries may have difficulty giving priority to provision of environmental services, especially when they face more economically attractive development options (Box 35).

This raises the question of the potential role of PES in poverty alleviation (FAO, 2007e). There are some

| BOX 35 | Willingness and ability to pay for conservation |

Conversion of biodiversity-rich delta to sugar-cane plantations

Kenya has recently embarked on a large-scale sugar plantation, converting about 2 000 km² of the pristine Tana River Delta, which provides habitat for a large number of species and a source of livelihood to local communities. The objections of conservationists and local communities have led to judiciary intervention, delaying project implementation.

Closure of sugar-cane plantations for wetland restoration

The United States Sugar Corporation, the largest producer of cane sugar in the United States of America, has agreed to close down about 750 km² of sugar-cane plantations to help in the restoration of the Everglades wetlands. The State of Florida will pay the company an estimated US$1.75 billion in order to purchase the land.

SOURCES: *Environment News Service*, 2008a, 2008b.

indications that the poor may not benefit particularly from ecosystem markets (FAO, 2004). The concern is to ensure that the payments for the environmental services actually go to the farmers who provide the services by adopting appropriate land use. However, their ability to provide the services depends largely on rights to and ownership of the land, as well as other policy and institutional factors that determine the transaction costs. Consequently, it is often the large landowners that are able to take advantage of PES arrangements.

An additional concern is that, given the social and economic inequities that exist in most countries, when markets develop and profits can be made by selling ecosystem services, the access of poor people to these services may be reduced.

A chief impediment to the provision of environmental services through existing approaches is the high transaction cost. Environmental markets are more sophisticated and complex than commodity markets, requiring substantial information on technical aspects of provision of services and well-developed institutional and legal arrangements. This again suggests the enormous effort required to develop effective measures to provide environmental services in most developing countries.

OUTLOOK

The outlook for the provision of environmental services is mixed. Growth in income coupled with greater awareness will usually strengthen demand for environmental services as well as the ability of a society to meet the costs of environmental protection. However, increased income often reduces environmental services as more goods and services are produced. In particular, countries with rapidly growing economies often go through a period when forest resources are exploited or converted to other uses, resulting in a decline in environmental services.

There is no single solution appropriate to all contexts. Both market and non-market approaches have their strengths and limitations. It is often assumed that economic growth is a prerequisite for improving the environment, but the reality is more complex. Many factors, including institutions and legal frameworks, will have an impact on the ability of a country to manage its forests in such a way as to provide stable or increasing environmental services.

Changing institutions

nstitutions are key to sustainable resource management and societal adaptation to social, economic and environmental changes. As in other sectors, the overall trend in forestry is towards a pluralistic institutional environment, attributable to two divergent trends: globalization and localization. Increased cross-border movement of capital, labour, technology and goods resulting from globalization has necessitated adaptation by existing institutions and the establishment of new ones. At the same time, local communities have become more involved in resource management through decentralization and devolution of responsibilities. While there are considerable differences across countries and regions, this chapter summarizes how institutions in the forest sector are responding to the emerging developments outlined in Part 1.

INSTITUTIONAL CHANGE IN FORESTRY: AN OVERVIEW

Before the 1990s, the forest sector was dominated by government forestry agencies, several large enterprises, a multitude of small and medium enterprises (many operating outside the formal system) and a few international organizations largely focused on providing technical support to public forestry agencies. Today, the forest sector is characterized by a greater number of institutions addressing a wider array of issues (Box 36).

The private sector and civil-society organizations have experienced significant growth since the 1990s, particularly as a result of:

- the political and economic changes following the collapse of the Soviet Union, especially the shift from centralized planning to market-oriented economic policies and globalization;
- growth in environmental awareness and concerns and the proliferation of related initiatives following the United Nations Conference on Environment and Development (UNCED) in 1992;
- changes in funding for forestry, i.e. increases in foreign direct investments and private-foundation

support (Box 37) alongside decreases in official development assistance.

Developments in information and communication technologies have further catalysed institutional changes, challenging hierarchical structures and calling for institutions to respond to the demands of a more informed public (see Box 45 on page 88).

BOX 36 | Types of institutions dealing with forest issues

Public forestry agencies and enterprises
- National policy formulation, legislation and planning, including national forest programmes
- Management of forests and forest industries and all related activities, including trade in forest products
- Regulatory and enforcement functions – providing a level playing field to other institutions involved in forest and tree resources management

Private sector
- Management of forests and other resources, including planted forests
- Production and processing of, and trade in, wood and non-wood products

Civil-society organizations
- Environmental and social advocacy in policy and market development and awareness generation

Informal sector
- Production and processing of, and trade in, wood and non-wood products

International and regional organizations and initiatives
- Intergovernmental forest policy, environment- and trade-related processes and conventions
- Financing, development and technical assistance, including technology transfer
- Regional collaboration arrangements
- Science and technology development and networking

PUBLIC AGENCIES

Government forestry agencies remain the dominant force in the sector. More than 80 percent of global forests are under public jurisdiction (FAO, 2006a).

Government forestry departments are often among the oldest of the civil services. Many originally focused on enforcing regulations, with the main objective of protecting and managing the forests to supply forest products and generate revenue for government. They traditionally integrated multiple functions from wood production to processing and trade as well as forestry research (see Box 38), education, training and extension.

Challenges of reduced public expenditure, mounting expectations of different stakeholders and increasing conflicts over the use of forest resources are stirring public agencies to rethink their management objectives, functions and structures (FAO, 2008h). The evolution in focus can be loosely described as moving from policing the forests to managing them to facilitating management by others (Table 26).

BOX 37	Growth of private foundations in the United States of America

The United States of America has the largest segment of private foundations supporting development activities. In 2005, they provided grants of an estimated US$3.8 billion (US$1.6 billion in 1998). Almost half of the support is in the field of health (largely because of the huge support provided by the Bill & Melinda Gates Foundation). Environment accounted for about 10.4 percent of the number of grants in 2004.

SOURCE: Renz and Atienza, 2006.

BOX 38	Declining public forestry research

In most countries, the public forestry administration has traditionally had a research arm, but institutional arrangements for research are changing. Research is increasingly carried out by government-funded independent organizations, universities and the private sector, often through collaborative networks. It is increasingly demand-driven rather than supply-driven. However, these changes raise concerns about imbalances in investments. Support for basic and strategic research has been declining, with more attention focused on applied and adaptive research that yields immediate returns. Furthermore, the results of private-sector research are often not publicly available.

TABLE 26
Progression in the development of public forestry agencies

Stage	Objectives of resource management	Functions and structures
Protecting	To utilize what grows under natural conditions (e.g. to log natural forests) To safeguard future timber supplies for strategic reasons	Policing of the forest estate Hierarchical structure
Managing	To improve the state of resources by investing in improved management To create assets, including planted forests	Production and resource management Emphasis on technical and managerial skills
Enabling others to manage	To support or empower other players (private sector, communities, farmers, etc.) to manage resources and regulatory functions	Creation of enabling conditions Negotiation, facilitation and conflict resolution Emphasis on diversity of skills and quick response to needs of various stakeholders

In some cases, reform has been superficial; for example, limited to changes in ministerial responsibility (in particular shifting between agriculture and environment ministries) or to structural but not functional change. Many public agencies are unable to develop the human resources needed in order to manage forest resources in an increasingly complex environment (Nair, 2004; Temu, 2004), and many lack sufficient capacity for long-term strategic planning or open sharing of information, with a tendency to be reactive to short-term pressures and concerns (often mirroring the larger public administration).

Strategies used in more successful transitions to an enabling role have included:
- separating policy and regulatory functions from management functions;
- entrusting wood production and processing to an independent commercial government entity or privatizing all commercial activities, usually as part of a larger policy of economic liberalization, often triggered by government budgetary crises (as in the case of New Zealand [O'Loughlin, 2008]);
- decentralizing and devolving management responsibility to the local level (Box 39), usually as part of a larger programme of political and administrative decentralization – with widely different results.

BOX 39	Elements of successful decentralized forest governance

External to the forest sector
- Significant transfer of power and responsibilities to democratically elected and accountable lower levels of government
- Fair and clear enforced property rights and an appropriate regulatory framework
- Respect of the law by governments, the private sector and civil society
- Effective linkages between government, the private sector and civil-society institutions

Internal to the forest sector
- Effective and balanced distribution of responsibilities and authority among different levels of government
- Adequate resources and institutional effectiveness at each level of government
- Sufficient participation of civil society and the private sector at all levels

SOURCE: Contreras-Hermosilla, Gregersen and White, 2008.

PRIVATE SECTOR

Private enterprises range from individual and household microenterprises and small farms, often operating on a minimal budget, to large transnational corporations, whose annual turnover in some cases exceeds the GDP of a small country.

Corporate sector

The corporate sector accounts for a large segment of logging concessions, planted forests and wood industries. Profitability remains its primary objective. Corporations operate in an extremely competitive environment with constant pressures to cut costs and improve market share. The following are some of the sector's major challenges and opportunities:
- Rapid growth of emerging economies in Asia is resulting in a regional shift in the demand for wood products (see the chapter "Global wood products demand" in Part 2). Investments in new capacity are taking place in countries where demand and profitability are perceived to be high and the costs of production – especially of fibre, energy and labour – are low. In particular, the pulp and paper industry has seen a spate of mergers and acquisitions and the closure of less-competitive mills.
- Pressure for industry to adhere to tenets of corporate social responsibility is expected to mount as society becomes more concerned about environmental and social issues (Box 40). "Green" values will influence procurement of goods and services along the whole supply chain. Consumer preference is shifting in favour of certified products, but this is not always reflected in higher prices.
- Climate change concerns are expected to provide new opportunities for wood products (which store carbon and require relatively little energy to produce) and industrial wood energy. Major related challenges include increasing transport costs owing to the rapid expansion of global value chains and increasing demand for wood.

Strategies for adaptation to the above challenges include:
- Focus on core business and divestment of non-core activities: The traditional model of large integrated industrial units is giving way to highly networked global supply chains, linking firms and affiliates across countries, including subcontractors and home workers operating outside the formal system. Components of production may be relocated abroad for improved profitability. Wood production may be outsourced to farmers through partnership arrangements. Forest product companies increasingly recognize that tying up large stocks of capital in forest ownership affects their short-run cash flow

| BOX 40 | Corporate social responsibility |

The overarching focus on profitability in the private sector often results in high social and environmental costs. As society's awareness of these costs increases, pressures mount on the private sector to abide by environmental and social regulations. Industry may also find it advantageous to project a green image, especially among environment-conscious consumers. Industry organizations have developed a number of criteria relating to corporate social responsibility, and green auditing is becoming mandatory. In Rome in 2006, chief executive officers from 61 companies belonging to the International Council of Forest and Paper Associations signed a commitment to global sustainability. The World Business Council for Sustainable Development has prepared guidelines on sustainable procurement of wood and paper products that address environmental and social aspects (WBCSD and WRI, 2007). Increasing environmental awareness and easy access to information will help to ensure that industry no longer neglects its responsibilities through superficial "greenwashing".

| BOX 41 | Institutional investors: TIMOs and REITs |

Most investments in planted forests have traditionally been made by government, smallholder or industrial forest owners. However, management arrangements such as timber investment management organizations (TIMOs) and real estate investment trusts (REITs) have created a significant shift in forest ownership from industries to institutional investors, primarily in North America but also in Australia, Finland, New Zealand, South Africa and Sweden. Investment by institutions in planted and managed native forests increased worldwide from less than US$1 billion in 1985 to more than US$30 billion in 2007. The number of TIMOs grew from two or three in the 1980s to more than 25 in 2007. About 20 million hectares of private forest land are under TIMO control. In the United States of America, forest landownership by integrated forest companies (those involved in both production and processing) declined from 19.5 million hectares in 1994 to 4 million hectares in 2007 (Neilson, 2007).

Some observers are concerned that the increase in forest ownership by wholly profit-seeking institutional investors could undermine long-term investments in forest management and research and also accelerate commercial development of private forest lands. However, the growth of TIMOs appears to be slackening because of the limited area available for sale.

SOURCES: FAO, 2007f; Sample, 2007.

| BOX 42 | Sovereign wealth funds: an emerging player in forestry investment |

Since 2001, foreign exchange reserves have grown rapidly, far beyond the established benchmarks of adequacy. The sovereign wealth fund (SWF) is a vehicle established by some governments to channel these reserves into investments. In the first quarter of 2008, the total assets held by 51 SWFs were estimated at US$3.5 trillion, and these assets are projected to grow to about US$5 trillion by 2010 and US$12 trillion by 2015. SWFs invest in many asset classes including real estate, plantations and government bonds. Four SWFs have already invested in timber lands.

SOURCES: FAO, 2007f; *Friday Offcuts*, 2008.

and stock market values. Divestment has led to the emergence of new players (Boxes 41 and 42).

- Investment in R&D: The corporate sector leads investment in R&D, focusing on applied and adaptive research and on the development of new products and processes to establish competitive advantage and to meet consumers' environmental demands. The sector often takes advantage of results from public research. Planted forests managed by the corporate sector are among the most productive.

Other private and community-based enterprises

Globalization provides new opportunities for small and medium enterprises, but they will need to adapt continuously to survive intensifying competition. Issues affecting the long-term performance of this vibrant institutional segment include:

- Ownership, legal framework and level playing field: Ownership and security of tenure are necessary for the development of any enterprise. Policies and legislation vary in the extent to which they provide land rights to local communities. In many countries, rules and regulations are crafted to the needs of large enterprises, leaving small and medium enterprises and community institutions at a disadvantage.
- Constraints on economic viability: Local communities often have access to only the most degraded and least productive land, which cannot provide benefits commensurate with the investments required. They often lack access to inputs (including credit) and markets. Many small enterprises focus on production of low-value-added products, which seldom help to enhance income. Local markets face increasing competition from global suppliers. The inadequacy of entrepreneurial skills to deal with changing opportunities and challenges remains the most critical constraint.

- Governance and distribution of benefits: In some local community enterprises, power imbalances lead to inequitable distribution of benefits, undermining long-term sustainability. This problem is particularly severe where democratic transparent systems of management and accountability are lacking and local vested interests dominate.

Factors that have helped small and medium enterprises cope successfully with the challenges include:
- improved access to information and opportunities created by the Internet, e-trading and other tools;
- upscaling of business activities through associations and federations and improvement of access to markets, inputs and services;
- increased efforts to develop technologies appropriate to the needs of small and medium enterprises;
- rapidly rising transport costs, making local value chains more competitive.

Stronger institutional arrangements are critical to the scaling up of operations and improved bargaining power. Moreover, they enable communities to take advantage of new technologies, which are vital to making community-based resource management economically viable.

CIVIL-SOCIETY ORGANIZATIONS

In recent decades, civil-society organizations have become major players in forest-related issues in most countries, often challenging established positions and increasing transparency. They have emerged as one of the main forces reshaping the future of forestry at all levels – local, national and global.

Indigenous peoples' groups have risen from local levels to become effective actors and advocates at the global level through coalitions presenting a unified front and delivering consistent messages in international meetings and processes. Their organized efforts have led to progress in recognizing and restoring the rights of indigenous peoples to forest land. The adoption in 2007 of the United Nations Declaration on the Rights of Indigenous Peoples, although non-binding, was a milestone.

Community forestry and community conservation organizations include: federations (Box 43); networks of local community organizations, advocacy and networking organizations, such as the Forest Peoples Programme; and coalitions, such as Friends of the Earth International, the World Rainforest Movement and the Global Forest Coalition. Reflecting the growth of community forestry around the world, these groups stress the connection between forests and livelihoods.

International environmental NGOs, such as the World Wide Fund For Nature, Conservation International, The Nature Conservancy, the Wildlife Conservation Society and the International Union for Conservation of Nature

---

BOX 43	A federation of forest communities in Nepal

The Federation of Community Forest Users Nepal (FECOFUN), a forest-user advocacy organization founded in 1995, provides national representation of local people's rights in resource management. Comprising rural farmers – men and women, old and young – from almost all of Nepal's 75 districts, FECOFUN exemplifies the evolution and maturation of a community-based group into an important rural institution. Indeed, it is Nepal's largest civil-society organization.

FECOFUN and community forestry in Nepal owe their success to recognition of rural people's dependence on forests and to institutional incentives structured in accordance with rural realities.

SOURCE: FECOFUN, 2006.

(IUCN) (an umbrella group of which all the others are members), are the most well-funded and perhaps the most effective civil-society actors in forestry. Although differing in perspectives and approach, these groups focus attention on conserving biological diversity, extending protected areas, driving forest certification and improving forest governance to reduce illegal logging and trade in endangered species.

A related group consists of civil-society organizations that promote market-based approaches to conservation and sustainable forest management, such as certification, fair trade, organic and sustainable agriculture, ecotourism and green investments. Some of these organizations, including FSC and PEFC, have brought about changes in the behaviour of producers and consumers of forest products.

A number of international environmental NGOs, for example the International Institute for Environment and Development (IIED) and the World Resources Institute (WRI), function as "think tanks", enhancing knowledge in key areas.

In addition, complex webs of national, regional and global networks, many still relatively informal, link farmers, forest-dependent communities, small traders and local activists. These alliances are no longer strictly the domain of large international conservation organizations and major development groups.

Overall, civil-society organizations form a strong counterforce against powerful players such as governments and the corporate sector. Their effectiveness stems largely from:
- close contact with grassroots and understanding of local issues;
- multidisciplinary approaches to resource management issues;

- effective communication with stakeholders and funding sources;
- skilful use of networks and associations and development of strong linkages with other players;
- their detailed research on key issues and its use in support of local action.

Increasing awareness and concern about social and environmental issues imply an increasing role for civil-society organizations in forestry.

The shift towards institutional and economic complexity should mirror more effectively the ecological and cultural diversity of forests and peoples. Such complexity is needed in order to help forests fulfil their integrating role in a dispersed, diversified and distributive forest economy. Civil-society actors inject much-needed disorder into intentionally neat power equations (J. Campbell, personal communication, 2008).

INFORMAL SECTOR

The dividing line between the formal and informal sectors is sometimes blurred, especially as many small and medium enterprises operate outside the formal realm. Players outside the formal sector range from traditional local forest management arrangements that have been pushed into the informal realm by restrictive government regulations to illegal logging networks that exploit weak institutional arrangements in many countries.

Although it is difficult to define the extent of its reach, the informal sector continues to be significant worldwide. The International Labour Organization estimates that for every job in the formal sector in forestry there is another one (or two) in the informal sector (ILO, 2001). Most of these are in production or collection of woodfuel and NWFPs. Unpaid subsistence work, primarily in woodfuel harvesting, is estimated to employ about 14 million workers (full-time equivalents), of whom 90 percent are in developing countries. Informal sector employment is often dominated by women.

Many small forest enterprises operate informally, largely because of ill-defined property rights and an unfavourable business environment with high barriers to entry and concomitant transaction costs. The informal sector dominates in countries where regulations are cumbersome and inflexible (World Bank, 2006). Increasing pressure by the formal sector to reduce costs is encouraging the growth of the informal sector. Work is often outsourced to firms outside the formal sector that cut production costs by failing to abide by social and environmental norms.

The key issue is whether governments will make significant efforts to create a favourable business environment by removing barriers restricting entrepreneurship. Improved access to credit, markets and technology could potentially move some players from the informal to the formal sector.

Also key are concerted efforts to address illegal logging, which currently include intergovernmental forest law enforcement and governance processes, tracking and verification systems and anti-money-laundering measures.

INTERNATIONAL ORGANIZATIONS

Although less quick to adapt than private-sector or civil-society organizations, international forest-related organizations have evolved in the past two decades. Before 1990, the United Nations (UN) and other intergovernmental organizations, international research and financing organizations and bilateral donor agencies provided mainly technical support, primarily focused on production of wood products. The priority areas were silviculture and forest management, forest industries, research, education, training and extension.

Since the UNCED in 1992, under the overarching objective of sustainable forest management, international organizations have broadened their agenda to address a wider array of social, economic and environmental issues. New types of international institutions have emerged (UN forest policy processes, environmental conventions and agreements and regional intergovernmental processes) and initiatives and partnerships have multiplied. Programmes place more emphasis on support to policies and institutions, with increased focus on governance, poverty alleviation and more recently on integrating forestry in the framework of the Millennium Development Goals. With growing concern about climate change, the pursuit of mitigation and adaptation measures is an emerging priority.

The proliferation of institutions and initiatives has necessitated substantial efforts to minimize fragmentation and avoid duplication. Duplication is a hazard because constituents commonly ask organizations to take up the latest "hot" issue – and organizations need to work where there is funding, which again tends to be available for these hot issues. Fragmentation at the international level accentuates problems at the country level, especially where development efforts are compartmentalized in different sectors. The capacity to coordinate is in short supply in countries where problems are most acute.

Efforts to address fragmentation and duplication include the "One UN" approach (UN, 2006b), which aims to coordinate the disparate activities of the various UN agencies at the country level, and the Collaborative Partnership on Forests (CPF) – an example of coordinated support for the international forest policy process (Box 44).

The accelerating pace of globalization and the emergence of a host of transboundary economic, social and environmental issues need to be addressed by effective international institutional arrangements. Some of the likely changes in the next few years may be:

- consolidation of institutions, in response to resource constraints and pressure to see concrete results on the ground;
- a shift from processes to tangible outputs and results, as demanded by a more informed society;
- increasing emphasis on regional, subregional and other group initiatives to enable countries with similar views and perceptions to address shared problems, and increased attention to forestry issues by regional and subregional economic blocks.

BOX 44	Collaborative Partnership on Forests

The Collaborative Partnership on Forests (CPF), a voluntary arrangement among 14 international organizations and secretariats with substantial programmes on forests, aims to enhance coordination of support to the international forest dialogue and to country-level implementation of sustainable forest management. Initiatives on streamlining forest-related reporting and harmonizing definitions have aided global, regional and national forest processes. Recent initiatives include a joint strategic response to the global climate change agenda and consolidation of scientific knowledge in support of international policy processes.

OUTLOOK

With the emergence of new players, the institutional landscape in the forest sector has become more complex and the balance among players is shifting. In general (although not in all countries), the playing field is more level, partly as a result of new information and communication technologies. The much-needed pluralism provides new opportunities for small and medium enterprises and community organizations. Civil-society institutions, usually focusing on social and environmental issues, and private-sector institutions, usually emphasizing economic aspects, are gaining in strength and number; funding and investments increasingly favour them over the public sector and international institutions. If the government forestry agencies that historically dominated the scene fail to adapt to these changes, they could fade into irrelevance. With the increasing pace of globalization, new players such as TIMOs, REITs, sovereign wealth funds and carbon trading institutions could alter the global institutional map. Institutions will face tremendous pressure to balance fragmentation and to consolidate efforts.

Views from CPF partners

The Non-Legally Binding Instrument and future priorities for forests

From the United Nations Forum on Forests (UNFF)

The Non-Legally Binding Instrument on All Types of Forests (NLBI) adopted by the UN General Assembly in December 2007 embodies a global consensus on sustainable forest management and outlines future priorities in the form of four Global Objectives on Forests:

- reverse the loss of forest cover worldwide through sustainable forest management;
- enhance forest-based economic, social and environmental benefits, including by improving the livelihoods of forest-dependent people;
- increase significantly the area of protected forests and other areas of sustainably managed forest worldwide;

- reverse the decline in official development assistance for sustainable forest management.

With the NLBI and its new multiyear programme of work, UNFF is poised to deliberate on some of the most pressing issues related to forests in the coming years. In 2009, UNFF will discuss the contribution of forests to addressing challenges of climate change as well as the role of forests in protecting biodiversity and reducing desertification. In this regard, issues such as governance and sound participatory decision-making will be crucial to ensuring that the benefits of forests are secured and that long-term planning takes precedence over short-term gains.

Developments in forest science and technology

The science and technology system involves basic and strategic research, applied and adaptive research, and adoption of the results. Broadly speaking, technology in forestry generally relates to two areas:

- management of forest and tree resources for the production of goods and provision of environmental services;
- harvesting, transport and processing of wood and non-wood products.

Within these areas, development tends to focus on one or more of the following objectives:

- reducing costs and increasing productivity;
- developing new products and services;
- conserving resources and reducing adverse environmental impacts;
- improving energy efficiency.

In meeting these objectives, newer fields, such as biotechnology, nanotechnology and information and communication technologies (Box 45), are having a notable impact.

A growing area of research deals with enhancing the scientific base for the provision of environmental services. This usually involves study of ecosystem processes and the implications of different degrees of human intervention. For

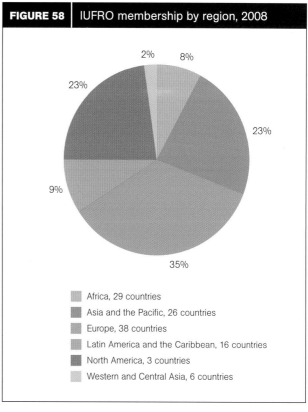

FIGURE 58 | IUFRO membership by region, 2008

- Africa, 29 countries
- Asia and the Pacific, 26 countries
- Europe, 38 countries
- Latin America and the Caribbean, 16 countries
- North America, 3 countries
- Western and Central Asia, 6 countries

SOURCE: IUFRO, 2008.

BOX 45 | Information and communication technologies in forestry

The growth of information and communication technologies (ICTs) has had significant direct and indirect impacts on forestry and has been central in accelerating the pace of globalization. The Internet and mobile communications have created unprecedented opportunities for those who were traditionally outside the global information loop, including small and medium-sized enterprises. ICTs have increased labour productivity, reduced costs and increased returns. Online stores provide marketing opportunities for wood product and service suppliers.

ICTs have also fostered institutional change in forestry. The increased ease of information sharing and global networking diminishes the power of vertically structured organizations and fosters the development of small organizations. ICTs have helped to promote transparency and accountability on an unprecedented scale, as very little information can be kept away from public access and scrutiny. ICTs have also facilitated awareness-raising about forest-related issues such as deforestation, biodiversity loss, forest fires and the marginalization of indigenous communities.

SOURCES: Hetemäki and Nilsson, 2005; Nyrud and Devine, 2005.

example, climate change mitigation and adaptation will require substantial efforts to understand carbon fluxes under different land uses and how ecosystems and species respond to changes.

Science and technology capacity differs significantly between developed and developing countries, mostly reflecting differing abilities to invest in education, training and infrastructure. Although a simplification, the geographical distribution of the members of the International Union of Forest Research Organizations (IUFRO) reflects the differences in research capacity among regions (Figure 58).

Translating scientific knowledge into technologies and then applying these remain major challenges, partly because of fragmented institutional arrangements. Adoption of technologies is context-specific. Often, a choice must be made among a wide array of technologies available for the same task.

Historically, public-sector forestry agencies led the development of forest science and technology. Today, there are many more players; in general, the public sector's role has diminished (Table 27) and its capacities have declined sharply in many countries.

SCIENCE AND TECHNOLOGY IN SELECTED AREAS

Forest management

For most of the twentieth century, natural forests were the main source of wood and other products, and forest research focused on managing them sustainably. Various silvicultural systems were developed (e.g. selection and shelterwood systems), taking into account the density of important species, their growth rates, light and moisture requirements, their ability to regenerate naturally and competition between marketable and non-marketable species. Low-intensity harvesting was adopted to avoid undermining forest environmental services. Vulnerable areas were excluded from logging.

Views from CPF partners

Research challenges of the future

From the International Union of Forest Research Organizations (IUFRO)

IUFRO is the central global network for forest researchers. It has three strategic goals:

- strengthen research for the benefit of forests and people;
- expand strategic partnerships and cooperation;
- strengthen communication and links within the scientific community and with students as well as with policy-makers and society at large.

To provide independent scientific expertise and information to its potential users, IUFRO periodically identifies emerging key issues and assesses its activities.

The forest research challenges of the future identified by IUFRO involve:

- the increased global demand for wood and non-wood goods and services;
- bioenergy;
- impacts of climate change;

- competition for land and how to reverse deforestation;
- the role of genetically modified trees and plantation forestry;
- invasions of alien pests and pathogens;
- biodiversity conservation;
- social and behavioural processes;
- the impact of global economic developments on local economies and livelihoods.

Society is placing greater pressure on scientists to explain their research publicly and to demonstrate its positive impacts. Moreover, the users of scientific information, including policy-makers and practitioners, increasingly want to be involved in the development of research agendas. Networking can help both the scientific community and its actual and potential beneficiaries to enhance research uptake and impact. ▪

TABLE 27

Key players in forest science and technology

Key players	Research focus	General trends
Public-sector forest research institutions	Basic and applied research in all aspects of forests and forestry A significant share of research is not demand-driven, but provides the foundation for downstream applied and adaptive research	With few exceptions, declining because of reduced funding and concomitant reduction in human resources Fragmentation of research agenda and weak linkages between research areas
Universities	Mostly focused on the science of forestry and to a limited extent on applied research leading to technology development	Declining public-sector funding compelling shifts in favour of more applied and adaptive research in collaboration with industry
Industry	Demand-driven research primarily undertaken by large enterprises Focused on applied and adaptive research leading to the development of new processes and products that can be patented	Increased investments to raise competitiveness Collaborative arrangements with public institutions and universities, largely to benefit from capacity in basic research
International public-sector research institutions and networks	Global and regional issues and research networking (but very few in number)	Shift in focus from technical aspects of forestry to policy issues, with increasing emphasis on social and environmental dimensions
Independent think tanks and civil-society research institutions	Mostly policy issues, with particular emphasis on environmental and social issues Focused on supporting advocacy initiatives	Expanding influence, especially in policy processes at national and international levels
Manufacturers of equipment and machinery	Production of machinery and equipment that draws on many technologies for specific tasks	Intense competition and the constant need to upgrade machinery and add new features

With a shift to the sourcing of wood from planted forests and the exclusion of large tracts of natural forests from wood production, these low-intensity management systems have been abandoned in many countries. The development of technologies that made it possible to process wood irrespective of its natural qualities and size has also contributed to shifting attention away from these systems.

Research in natural forests now focuses more on integrating environmental, social and economic objectives according to the principles of sustainable forest management. A number of national, regional and international initiatives focus on the development of criteria and indicators for measuring progress towards sustainable forest management, outlining the nature of technology to be adopted. Implementation of sustainable forest management requires substantial strengthening of the science and technology base. To this end, much research focuses on ecosystem structure and functioning, the spatial and temporal linkages among ecosystem components and processes, and their relation to the immediate and larger social and economic context. However, implementation of such research remains a challenge (CIFOR, 2004), especially in developing countries.

Technologies that increase the speed at which vast amounts of spatial and temporal data can be analysed and synthesized are revolutionizing forest management in some countries and are expected to spread. Improvements in the resolution of satellite imagery and the development of software to interpret it will contribute to real-time monitoring of deforestation, pests and diseases, fires and other potentially devastating events. Geographic information systems (GISs) and global navigation satellite systems provide forest managers with increasingly precise information on the nature and condition of forest resources, which can be processed and transmitted rapidly (Box 46). This information is also valuable as evidence for public consultation, verification of legality and third-party certification.

New modelling and visualization software linking GISs and remote sensing provides high-quality digital simulations of future forest landscapes to reveal changes that might result from natural processes, such as climate alterations, or human interventions, such as planting, thinning and harvesting. Such simulations facilitate community engagement in forest management decision-making (Sheppard and Meitner, 2005).

An increasingly urgent area of research relates to the adaptation of forests to climate change. For example, genetic and environmental variation in tree growth and health is being studied in order to predict potential impacts of climate change on ecosystems and species ranges, to predict adaptive responses of tree populations to climate and to formulate new strategies to help forest trees adapt to the changing climate (Wang et al., 2008).

Planted forests and wood production

Planted forests have received most investment in forestry, and also in forestry technology development. Research aims primarily to enhance productivity through faster growth rates and to improve wood quality and the ability

BOX 46 | Remote-sensing applications in forestry

Remote-sensing techniques (including aerial photography and satellite imagery) have been used successfully for forest mapping and monitoring and make it possible to cover large areas consistently and cost-effectively. New technologies address technical challenges such as the variable height, structure, density and composition of forests. Airborne light detection and ranging using lasers can provide highly accurate estimates of tree cover and height; it can even assess the shape of individual trees. Space-borne radar (radio detection and ranging) is a promising new way to obtain estimates of stand volume and biomass and can penetrate clouds, overcoming some of the limitations of optical satellite sensors. New spectral sensing systems can measure a wide array of land and vegetation characteristics, making it possible to assess a range of forest attributes – helping to improve mapping of forest pests and diseases, for example.

SOURCE: R. Keenan, personal communication, 2008.

of forests to withstand adverse environmental conditions, pests, diseases and other hazards.

Enormous productivity increases have been obtained for fast-growing, short-rotation species such as eucalypts, tropical pines and poplars. For example, eucalypt plantings in Brazil have reached productivity levels exceeding 50 m³ per hectare. Productivity increases have primarily been a consequence of the cumulative impact of improved planting material, nursery practices, site/species matching and intensive site management. Substantial efforts have also been directed at improving the quality of management, for example through integrated pest management.

The focus on short-rotation, fast-growing species is directly related to demand from processing industries (producing pulp and paper and reconstituted fibreboard). Industry has been one of the main drivers encouraging innovation in wood production technologies. The new developments are mainly applied by the corporate sector – which, however, only accounted for about 18 percent of the world's productive planted forests in 2005. Thus, governments and smallholders (which hold 50 and 32 percent of planted forests, respectively) have not been able to adopt many of the improved technologies, suggesting considerable scope for enhancing productivity on the global scale.

Tree improvement programmes aim to accelerate the development and mass multiplication of progeny with desirable characteristics. Molecular techniques make it possible to characterize genetic diversity in trees,

insects and soil and plant microbes. While traditional improvement techniques rely on natural genetic variation, increasing but controversial efforts are also under way to develop genetically modified trees (Box 47).

Completed genome mapping of *Populus trichocarpa* has enhanced the understanding of genetic functioning in trees. A recently initiated effort to map the genome of *Eucalyptus grandis* (International *Eucalyptus* Genome Network, 2007) will further develop this capacity. Forest biotechnology can also improve the knowledge of cell function, allowing a greater understanding of processes such as wood formation, stress tolerance and carbon fixation and sequestration.

Soil and water depletion and biodiversity loss are other issues raised in the context of planted forest expansion. FAO's voluntary guidelines for the responsible management of planted forests (FAO, 2006f) propose a holistic approach that gives due attention to economic, social and environmental dimensions.

Agroforestry

Research on agroforestry, which comprises varied practices integrating crops, livestock and trees, aims to optimize these components in order to meet the economic, social, cultural and environmental needs of communities and households, while taking advantage of site-level variation in soils, topography and light and moisture availability.

Agroforestry technologies are generally ecologically and culturally site-specific. They have traditionally been developed through "hands-on" experience and transmitted through the generations. Successful agroforestry systems and practices include alley cropping, silvipasture, windbreaks,

BOX 47 | Genetically modified trees: blessing or curse?

Advances in gene transfer technologies and tree genomics are providing new avenues for genetic modification of trees. Traits considered for genetic modification include herbicide tolerance, reduced flowering or sterility, insect resistance, wood chemistry (especially lower lignin content) and fibre quality, which could all boost economic potential. Increasing interest in cellulosic biofuels is focusing greater attention on genetic modification, in particular on reducing the lignin content in wood. However, research and deployment, including field trials of genetically modified trees, remain a contentious issue. Concerns have been raised about impacts on ecosystems, especially potential invasiveness, impacts on biodiversity and the transfer of genes to other organisms.

SOURCES: Evans and Turnbull, 2004; FAO, 2006f.

hedgerow intercropping, parklands, home gardens and relay cropping. Some have been in existence for centuries, evolving in response to needs and constraints both on and off the farm. Formal agroforestry research applies the tools and techniques of modern science to help improve the traditional practices and enable their wider application. It generally takes a holistic perspective in that economic and other benefits are assessed with consideration given to the links among the different components.

Agroforestry is currently responding to new market opportunities. Planting of trees on farms to supply wood to forest industries has increased significantly in many countries. Accordingly, new research issues have emerged, including for example interactions between tree crops and food crops and long-term sustainability of production with a focus on maintaining and improving productivity of land.

Harvesting and processing of wood products

Improving economic efficiency and minimizing environmental damage have been the primary objectives of harvesting innovations. Shortages and increasing costs of labour have encouraged significant mechanization of logging and transport. Sophisticated harvesting, conversion and transportation technologies have been deployed in a number of countries, especially in industrial forest plantations.

Reduced-impact logging was developed in response to concerns about the long-term sustainability of wood production from natural forests. It involves measures to minimize damage to the remaining vegetation, enabling rapid recovery after logging. FAO has developed global and regional codes for sustainable forest harvesting and supports countries in developing national codes and guidelines. While the importance of reduced-impact logging is understood and its long-term commercial feasibility has been demonstrated, its adoption depends on the objectives of the resource owners or logging concessionaires and their willingness and ability to comply with market and non-market signals.

New techniques have been developed to identify the source of logs using tags, paints and chemical compounds that can be read by detection devices. New-generation radio-frequency identification tags and bar codes can easily track the movement of logs from forests to markets, helping to distinguish legally from illegally sourced wood.

Technological developments in wood processing largely focus on:

- economic competitiveness, with an emphasis on reducing costs, improving quality and developing new products;
- energy efficiency and production of energy during wood processing;
- compliance with environmental standards, for example by reducing effluents and reusing water

through "closed-loop processing" in the pulp and paper industry (Natural Resources Canada, 2008b).

Many technological developments in wood processing have been consumer driven, as processing is near the end of the forest products value chain, close to consumers and, thus, compelled to respond to changing demands. Intense competition has also encouraged innovation.

Traditional wood use was largely based on physical properties, especially strength, durability, working quality and appearance. Wood-processing technologies have improved mechanical and chemical properties, expanding uses and making it possible to employ species that were once considered less useful – for example, to use rubberwood (*Hevea brasiliensis*) for furniture and medium-density fibreboard. Biotechnology in the wood products sector has the potential to improve wood preservation properties.

New sawmilling technologies include laser and X-ray scanners combined with high-power computing, which make it possible to scan and store information on log diameter, length and shape and to produce optimal sawing patterns for each log to maximize sawnwood recovery (Bowe *et al.*, 2002). Picture analysis to determine surface properties (e.g. knots and colour) has improved the sorting and grading of sawnwood. New methods have been introduced to control the drying process and to measure physical strength, revealing possible defects (Baudin *et al.*, 2005).

Other technological developments in wood processing include:

- improved rate of recovery and the use of small-dimension timber, largely through improvements in sawmilling technologies and production of sliced veneer and reconstituted panels;
- recycling, for example use of recovered paper;
- the use of micro-organisms to bleach pulp and treat effluents in the paper industry, reducing costs and environmental impacts;
- total use of wood through biorefineries producing a range of biomaterials and energy (Box 48).

Nanotechnology, defined as the manipulation of materials measuring less than 100 nanometres (with 1 nanometre equalling one-billionth of a metre), is expected to revolutionize all aspects of production and processing, from production of raw materials to composite and paper products, permitting major advances in energy and material efficiency (Roughley, 2005; Reitzer, 2007). Most leading wood-product-producing countries are working on nanotechnology applications. Potential uses (Beecher, 2007) include:

- lighter-weight stronger products developed from nanofibres;
- coatings to improve surface qualities;
- production that uses less material and less energy;

- "intelligent" products with nanosensors for measuring forces, loads, moisture levels, temperatures, etc.

Non-wood forest products

NWFPs are diverse and many different technologies are used in their production and processing. Although most NWFPs are subsistence products, collected from the wild and consumed locally with minimal processing, some have been domesticated and are cultivated and processed using sophisticated technologies to meet the demand from global markets. Science and technology development for these products has focused largely on more organized systems of production, while subsistence production has relied almost entirely on indigenous knowledge.

Natural-resource degradation coupled with increasing demand has been the main driver of organized cultivation of many NWFP-yielding species – much as wood production has shifted from natural to planted forests. Research on domestication and cultivation has also been encouraged by the complexity and uncertainty of managing production in the wild. For many products, such as natural rubber, rattan, bamboo and some medicinal and aromatic plants, organized production and chemical substitution of natural components have virtually replaced collection from the wild, except for products intended for niche markets paying a high premium.

Scientific research has focused on:
- understanding the composition, properties and potential uses of different products;
- low-cost technologies for the extraction and isolation of marketable components and for the addition of desirable characteristics, e.g. to facilitate storage and transportation;

- improving processing technologies and developing new products, e.g. new plant-based pharmaceuticals and health and beauty products (the areas where most technological advances are taking place).

Technological developments, for example in biotechnology, present new opportunities and challenges for many NWFPs. While new uses and markets have emerged, so have substitute products that undermine existing markets. Petrochemicals and new technologies for processing glass and metals have significantly changed the markets for a number of plant-based products. NWFPs with limited end uses are particularly vulnerable to such developments. In contrast, bamboo has been developed for diverse end uses and has become a widely distributed material and an important source of income (FAO, 2007g).

Wood for energy

Woodfuel is (and is likely to remain) the main source of domestic energy for cooking and heating in most developing countries. Although increasing income and availability of more convenient fossil fuels have reduced wood energy use, this situation seems to be changing as a result of high fuel prices, perceived risks of fossil fuel dependence and growing concern about greenhouse gas emissions from the use of fossil fuels (FAO, 2008d).

Traditional wood energy systems rely on low-cost technologies affordable to low-income consumers. The technologies used vary in cost and in production and conversion efficiency. For example, charcoal is produced using a range of kiln types, from traditional mud to metal. Modern wood energy production using cofiring (combustion of biomass together with other fuel such as coal) or wood pellets involves considerably higher investments, but is also much more energy-efficient.

Substantial investments are being made to develop and commercialize technologies for producing biofuel from cellulose. How cellulosic biofuel will develop depends on its cost-competitiveness with fossil fuels and other alternatives. If high energy prices persist, cellulosic biofuel production is expected to become a major source of commercial energy. The impact on the forest sector remains uncertain, especially considering that other high-productivity feedstocks could be used rather than wood (e.g. switchgrass, *Panicum virgatum*).

Provision of environmental services

Scientific knowledge is essential for timely and appropriate decision-making to ensure the provision of environmental services by forests. As this knowledge is frequently incomplete, enhancing it needs to be a priority area for research. Examples include: the limited information on the economic consequences of changes in ecosystem services; the lack of quantitative models linking ecosystem change to environmental services; and the poor understanding of ecosystem structure and dynamics that determine thresholds and irreversible changes.

Breakthroughs will be necessary in order to address the drastic degradation of dryland ecosystems, which will be aggravated by decreased rainfall expected as a consequence of climate change. Many affected countries do not have the capacity to undertake the scientific programmes required, and international support will be necessary.

Natural and planted forests offer significant greenhouse gas mitigation potential. However, there are large gaps in knowledge of the role of trees and forest ecosystems in climate change processes and the effect of changes in forest cover on forest carbon stocks and greenhouse gas emissions.

Research on the protective role of coastal forests has intensified since the December 2004 tsunami in Southeast Asia but is still not conclusive. In more than 20 studies carried out in the two years following the tsunami, some researchers found that coastal forests reduce adverse impacts significantly, while others discovered that forests can also be a liability by adding to the debris that can damage human settlements (FAO, 2007h).

Forest hydrology research addresses areas such as the relationship between land use and water yield, an area where myths and misconceptions often dominate decision-making.

Because of the complexity and breadth of issues involved in non-marketed environmental services, it is difficult for scientists to influence direct drivers of change – policy-makers and development actors – in their decisions and practices (and to gain their support for research activities to obtain new relevant knowledge). However, the Intergovernmental Panel on Climate Change (IPCC) has shown that concerted, holistic scientific efforts at the global level can effectively raise awareness and improve understanding of important complex issues, identify key areas where uncertainties need to be reduced and support the research activities needed to make this happen.

INDIGENOUS KNOWLEDGE

The advances in modern science and technology outlined above have had significant impacts on the forest sector. However, for vast populations these technologies remain inaccessible. Many continue to depend on indigenous or traditional knowledge in managing forests and other natural resources (Parrotta and Agnoletti, 2007). Traditional knowledge is defined as "a cumulative body of knowledge, practice and belief, handed down through generations by cultural transmission and evolving by adaptive processes, about the relationships of living beings (including humans) with one another and with their forest environment" (UNFF, 2004). Such knowledge, developed long before the advent of formal forest science, is the mainstay of many forestry practices (Asia Forest Network, 2008).

Indigenous knowledge is of growing interest to forest science as it is increasingly recognized that indigenous resource-management systems can help to improve the

Views from CPF partners

CIFOR's new strategy: a focus on climate change
From the Center for International Forestry Research (CIFOR)

CIFOR has a vision of a world in which forests remain high on the world's political agenda and people recognize the real value of forests for maintaining livelihoods and ecosystems services. In this vision, decision-making that affects forests is based on solid science and reflects the perspectives of developing countries and forest-dependent people.

Stakeholders surveyed for input to CIFOR's new strategy for 2008–2018 cited climate change as the most significant forest-related environment and development challenge today, followed by forest governance, deforestation and the impact of fast-growing economies on forests (CIFOR, 2008b). Thus, CIFOR's research agenda focuses on six domains:

- enhancing the role of forests in climate mitigation (with a focus on reducing emissions from deforestation and forest degradation);

- enhancing the role of forests in adaptation to climate change;
- improving livelihoods through smallholder and community forestry;
- managing trade-offs between conservation and development at landscape scale;
- managing the impacts of globalized trade and investment on forests and forest communities;
- sustainable management of tropical production forests.

An additional cross-cutting theme addresses the gap between changing societal demands from the forest sector and current institutional arrangements and capacities.

In analysing issues and communicating results, CIFOR will include the perspectives of less powerful stakeholders such as women, forest-dependent communities and developing countries. ▪

framework for sustainable forest management. Low-input traditional land-use practices are particularly attractive in the context of declining energy supplies and increasing impacts of climate change. Traditional knowledge provides alternatives to modern science, especially in health care. For example, South Asian Ayurveda and Chinese indigenous medicine are increasingly practised throughout the world, and the use of plant-based pharmaceuticals is growing rapidly.

In efforts to improve the livelihoods of poor marginalized indigenous communities, it is essential to understand their traditional knowledge – their values, perceptions and knowledge of their local ecological conditions. With social, economic, political and institutional change, indigenous knowledge provides opportunities but also faces challenges (Box 49). Several scenarios are evolving:

- Domination, marginalization and assimilation: Despite increasing recognition of their rights, indigenous people are systematically marginalized in many countries, including by narrowly focused development programmes. As vast tracts of forests that sustained indigenous communities are converted to other uses, forest-based livelihoods and the associated knowledge are soon lost.
- Selective appropriation: Realization of the economic potential of traditional knowledge (particularly in the rapidly expanding pharmaceuticals and health and

beauty care markets) has led to systematic efforts to identify and commercialize it – taking the knowledge out of its social and cultural context and raising issues of intellectual property rights and fair compensation for knowledge holders.
- Rediscovery: Increasing emphasis on protecting the rights, cultures and technologies of indigenous communities can create a favourable environment for the natural evolution of traditional knowledge. Developments in the international policy arena, such as the passage of the UN Declaration on the Rights of Indigenous Peoples, specifically recognize the need to respect traditional knowledge and practices.

Indigenous knowledge and community-based innovation are dynamic. Options for action include creating incentives to improve the capacity of formal research organizations to work with local and indigenous people and encouraging collaboration in conservation (IAASTD, 2008).

OUTLOOK

Visualizing the future of forest science and technology is difficult in a context of rapid change. Innovation has significantly improved the capacity of the forest sector to meet the changing demands of society and will continue to do so. However, many developing countries have little or no credible science capacity, and this lack hinders

---

BOX 49	Strengths, weaknesses, opportunities and threats for the survival of traditional forest knowledge

Strengths
- Adapted to local environmental, social, economic and cultural context
- Holistic, with focus on community welfare
- Integrated, avoiding artificial barriers of formal scientific disciplines
- Less resource-demanding and consequently more sustainable

Weaknesses
- Often not codified or widely disseminated – hence, not easily transferred and vulnerable to erosion over time
- Inadequately nurtured and developed
- Limited in ability to meet demands of increasing populations or large areas

Opportunities
- Increasing focus on sustainable management of resources adapted to local conditions and emphasizing social, environmental and cultural dimensions
- Emergence of pluralistic institutional arrangements and increasing emphasis on local community empowerment

- Increased interest in cherishing cultural diversity and growing niche market for unique products and services
- New information and communication technologies improving interaction and collaboration among indigenous groups

Threats
- Globalization and mass production undermining markets for goods and services produced locally using indigenous knowledge
- Marginalization and impoverishment of indigenous communities through appropriation of their land and other resources and consequent loss of culture and knowledge
- Ill-defined rights permitting appropriation of knowledge for commercial interests (bioprospecting) without appropriate compensation
- High investment in mainstream science and technology overshadowing traditional knowledge

their long-term developmental potential. Even in many developed countries, forest science and technology capacity has eroded.

The growth of commercially driven private-sector research and the declining capacity of public-sector research raise a number of issues. Most private-sector efforts are driven by the objective of maintaining competitiveness. As a result, it is often restricted in access, it may neglect environmental and social dimensions and it does not tend to nurture more open-ended upstream basic research. Vast populations that cannot afford to pay for improved technologies are excluded from the benefits. This accentuates disparities in access to knowledge, with consequences for income and living standards.

More concerted efforts are needed to address the imbalances and deficiencies in scientific and technological capacity. Challenges to governments include:

- reducing barriers to the flow of technologies among and within countries;
- ensuring that social and environmental issues are mainstreamed;
- transcending traditional sectoral boundaries to take advantage of science and technology developments outside the forest sector;
- setting a clear policy framework indicating the objectives, priorities and strategies for developing forest science and technology.

Finally, while this chapter has addressed biophysical aspects of forest science, the study of human behaviour, including economics and sociology, is equally important. Countries need to address both areas in a balanced way. Indeed, inadequate attention to the social-science dimension may be one of the reasons for the weak links between science and policy in many countries.

Postscript – challenges and opportunities in turbulent times

As *State of the World's Forests 2009* goes to press (late 2008), the world is experiencing a steep economic decline. The contraction of the housing sector and the subprime mortgage crisis in the United States of America have severely affected financial markets, triggering a global economic slowdown and recession in several countries. Confidence in financial institutions has eroded significantly. Stock market declines have reduced asset values by hundreds of billions of dollars. Deleveraging by banks seeking to secure their capital base has resulted in a credit squeeze that has affected all economic activities. A downward spiral has ensued, with decreases in production, employment, incomes and consumer demand causing further curtailment of production and further economic decline.

This downturn has affected almost all countries and transformed previously upbeat economic forecasts (IMF, 2008; UN, 2009). Global unemployment is expected to increase by 20 million in 2008 and 2009, potentially reversing recent success in poverty reduction (ILO, 2008a). Wages are expected to decline significantly (ILO, 2008b). Economic slowdown in most developed economies has already had consequences for emerging and developing economies, especially those dependent on exports and foreign direct investment. Official development assistance and remittances by migrant workers are expected to drop substantially (Cali, Massa and te Velde, 2008).

As an integral part of the larger economy, the forest sector will be affected by the overall economic slump. The severity of impacts will vary across the sector depending on linkages with sectors directly affected by the crisis.

DECLINING DEMAND FOR WOOD PRODUCTS AND SCALING DOWN OF PRODUCTION
The collapse of the housing sector, which has been at the epicentre of the current crisis, is a major blow to wood industries. The annual rate of new housing starts in the United States of America declined from about 2.1 million in early 2006 to less than 0.8 million in October 2008 (see figure on right). Several other countries, especially in Western Europe, have witnessed similar declines in

the housing sector, although not of the same magnitude. The housing decline has led to decreases in wood demand (UNECE and FAO, 2008; WWPA, 2008). Wood fibre demand in North America alone is expected to fall by more than 20 million tonnes in 2009 (RISI, 2008). Consequently, scaling down of production is widespread in almost all countries and all forest industries, from logging to sawmilling to production of wood panels, pulp, paper and furniture. Countries that are highly dependent on United States markets, for example Brazil and Canada, have already been severely affected.

Declining demand for forest products and the credit crunch together are having a severe negative impact on new investments, affecting all wood industries. As existing facilities remain underused or close down, investments in new capacities are being deferred or dropped.

REDUCED WILLINGNESS TO PAY FOR ENVIRONMENTAL SERVICES
The economic crisis could affect the demand for environmental services, especially as society becomes less able to pay. National and international policies coupled with a nascent market mechanism form the foundation for the growth in demand and supply of environmental services.

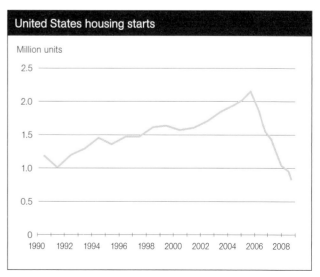

United States housing starts

Million units

SOURCE: NAHB, 2008.

Prolonged economic stagnation could have a negative impact on a number of environmental services unless the building up of a "green economy" is seen as a way out of the crisis.

Despite an initial appearance of stability, carbon markets have also been hit by the financial crisis as it has led to the collapse of some of the major investment banks involved in carbon trading. Carbon prices have plummeted in line with the prices of oil and other commodities. On the European Climate Exchange, carbon prices dropped from about €29 per tonne in early July 2008 to about €15 per tonne in mid-November 2008. An economic slowdown implies a decline in emissions from industries and power utilities, reducing the demand for emission allowances. Unless the price of carbon increases significantly and remains stable, the market approach to combating climate change could become ineffective. Its viability will largely depend on economic recovery and strong political commitment to conclude the post-Kyoto climate change agreement.

A more general concern is that some governments may dilute previously ambitious green goals or defer key policy decisions related to future climate change mitigation and adaptation as they focus on reversing the economic downturn (Egenhofer, 2008; Rice-Oxley, 2008). For example, commitment to European legislation on climate change, especially on auctioning emission allowances, is meeting obstacles, although some countries (e.g. the United Kingdom) have moved ahead with partial auctioning. Initiatives such as those for reducing emissions from deforestation and forest degradation (REDD) that are dependent on international financial transfers could face similar problems.

The unprecedented investment boom in biofuel production of the past few years is also fading. The slowdown could particularly affect investment in the more efficient second- and third-generation technologies, including lignocellulosic biofuel production.

Travel and tourism, including ecotourism, is another sector already affected by the economic slump. Since mid-2008, the expansion of international tourism has decelerated, initially because of high fuel prices and subsequently because of slowed economic growth and consequent lower consumer expenditure on travel and tourism (WTO, 2008). An already visible decline in international tourist arrivals in Kenya, South Africa and the United Republic of Tanzania, for example, heralds coming difficulties for wildlife tourism.

IMPACTS ON FORESTS AND FOREST MANAGEMENT: BAD NEWS AND GOOD

Reduced wood demand could have positive effects on forest resources, but the economic crisis could also reduce investment in sustainable forest management and favour illegal logging. Contraction of formal economic sectors often opens opportunities for expansion of the informal sector, including illegal logging. For example, a number of countries in Southeast Asia witnessed an increase in illegal logging following the 1997/98 economic crisis (Pagiola, 2004). Declining demand for high-priced wood from legal operations, reduced institutional capacity to protect forests as a result of lower budgets and increasing unemployment in the formal sector could increase illegal logging.

As outlined in the preceding chapters, rapid economic growth and reduced land dependence have helped to slow forest clearance and even reverse deforestation in many countries in the past decade. In several countries, remittances from migrant workers have helped to diminish the pressure on land. A continuing economic crisis could reverse the decline in dependence on agriculture, especially as industrial and services sectors decline and remittances drop. Increasing unemployment in the latter sectors could result in the return of workers to rural areas with attendant impacts on land use, including expansion of cultivation on to forest land.

Although smallholder cultivation may expand, large-scale cultivation of commercial crops, which has been a key driver of deforestation in the tropics, could decline substantially with the credit squeeze and the reduction in demand caused by the economic slowdown. Prices of palm oil, rubber and soybeans have dropped dramatically in the second half of 2008 (see figure on next page). While this is bad news for the producers of these commodities, it could be good news for forests. For example, the price of soybeans has a direct correlation with forest clearance in the Amazon basin (Nepstad et al., 2008).

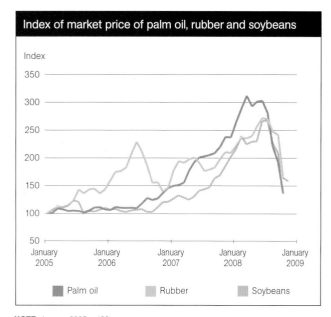

Index of market price of palm oil, rubber and soybeans

Index

350

300

250

200

150

100

50

January 2005 · January 2006 · January 2007 · January 2008 · January 2009

Palm oil   Rubber   Soybeans

NOTE: January 2005 = 100.
SOURCES: FAO, 2008; Index Mundi, 2008.

WEATHERING THE ECONOMIC STORM

Governments and central banks have acted rapidly to counter the crisis in a coordinated manner. However, nobody can be certain when the decline will bottom or how long it will take for markets and consumer confidence to turn around again. A recovery within a couple of years is a highly optimistic scenario. Many economists visualize further decline before a prolonged period of slow recovery. Wood demand is unlikely to reach the peak of 2005–2006 again in the foreseeable future.

Almost all countries are currently implementing monetary and fiscal policies to boost credit availability, growth and consumer demand. The forest sector could seize the opportunity to play a part in these fiscal stimulus measures – through the building up of natural resource capital (e.g. through afforestation and reforestation and increased investment in sustainable forest management), generation of rural employment and active promotion of wood in green building practices and renewable energy.

Economic cycles also always present opportunities for industry restructuring. Large-scale enterprises frequently rationalize production capacity by closing down old and inefficient units and focusing on the more productive part of the business. Large enterprises often face the greatest problems in an economic downturn; small and medium-sized enterprises may even find that the crisis offers them some opportunities.

The forest sector's ability to take advantage of the windows of opportunity provided by the current economic crisis will largely depend on institutional reinvention (see the chapter "Changing institutions" beginning on page 80). Difficult as this is, the crisis may stimulate acceptance and implementation of long-overdue reforms.

REFERENCES

Cali, M., Massa, I. & te Velde, D.W. 2008. *The global financial crisis: financial flows to developing countries set to fall by one quarter.* London, Overseas Development Institute.

Egenhofer, C. 2008. *Climate change policy after the financial crisis: the latest excuse for a new round of state aid?* CEPS commentary (available at www.ceps.eu).

FAO. 2008. International commodity prices (available at www.fao.org/es/esc/prices).

ILO. 2008a. *ILO says global financial crisis to increase unemployment by 20 million.* Press release ILO/08/45, 16 October. Geneva, Switzerland, International Labour Organization (also available at www.ilo.org/global/About_the_ILO/Media_and_public_information/lang--en/index.htm).

ILO. 2008b. *Global wage report 2008/2009.* Geneva, Switzerland.

IMF. 2008. *Global economic outlook.* Washington, DC, International Monetary Fund.

Index Mundi. 2008. Commodity price indices: rubber monthly price (available at www.indexmundi.com/commodities/?commodity=rubber).

NAHB. 2008. Housing starts. National Association of Home Builders (available at www.nahb.org/generic.aspx?genericContentID=45409).

Nepstad, D.C., Stickler, C.M., Soares-Filho, B. & Merry, F. 2008. Interactions among Amazon land use, forests and climate: prospects for a near-term tipping point. *Philosophical Transactions of the Royal Society,* 363: 1737–1746.

Pagiola, S. 2004. *Deforestation and land use changes induced by the East Asian economic crisis.* EASES Discussion Paper Series. Washington, DC, World Bank (also available at ideas.repec.org/p/wpa/wuwpot/0405006.html).

Rice-Oxley, M. 2008. Financial crisis threatens climate-change momentum. *Christian Science Monitor,* 13 November (available at features.csmonitor.com/environment).

RISI. 2008. *RISI's International Woodfiber Report predicts North American woodfiber demand to fall more than 20 million tons by year-end.* Press release, 23 October. Resource Information Systems Inc. (available at www.risiinfo.com/pages/abo//news/2008/2008-10-23.jsp).

UN. 2009. *World Economic Situation and Prospects 2009 – Global Outlook 2009.* Pre-release. New York, USA, United Nations (also available at www.un.org/esa/policy/wess/wesp.html).

UNECE & FAO. 2008. *Forest Products Annual Market Review 2007–2008.* Geneva, Switzerland, United Nations Publications.

WTO. 2008. *Slowdown in tourism growth reflects current uncertainties.* Press release, 10 November. World Tourism Organization (available at www.unwto.org/media).

WWPA. 2008. *U.S. financial crisis will delay recovery of housing, lumber markets until 2010.* Western Wood Products Association (available at www2.wwpa.org/Portals/9/docs/r-2008-09%20forecast.doc).

Annex

Notes on the annex tables

In all tables, the regional breakdown reflects geographical rather than economic or political groupings.
– = not available
0 = either a true zero or an insignificant value (less than half a unit)

In Table 1, "land area" refers to the total area of a country, excluding areas under inland water bodies. The world total corresponds to the sum of the reporting units; about 35 million hectares of land in Antarctica, some Arctic and Antarctic islands and some other minor islands are not included. Per capita gross domestic product (GDP) is expressed at purchasing power parity (PPP).

In Tables 2 and 3, data for Serbia and Montenegro are reported together as only combined data are available.

In Table 3, "biomass" refers to above-ground and below-ground biomass. Totals and subtotals refer only to those countries that have reported data on growing stock, biomass and carbon stock.

In Table 6, employment is reported for the formal forestry sector only.

TABLE 1
Basic data on countries and areas

Country/area	Land area	Population 2006				GDP 2006	
		Total	Density	Annual growth rate	Rural	Per capita (PPP)	Annual real growth rate
	(1 000 ha)	(1 000)	(population/km²)	(%)	(% of total)	(US$)	(%)
Burundi	2 568	8 173	318	4.0	89.7	333	5.1
Cameroon	46 540	18 174	39	2.1	44.5	2 089	3.8
Central African Republic	62 300	4 264	7	1.7	61.8	6 960	4.1
Chad	125 920	10 468	8	3.2	74.2	1 478	0.5
Congo	34 150	3 689	11	2.2	39.4	3 487	6.4
Democratic Republic of the Congo	226 705	60 643	27	3.2	67.3	281	5.1
Equatorial Guinea	2 805	495	18	2.3	60.9	27 161	−5.6
Gabon	25 767	1 310	5	1.6	15.9	14 208	1.2
Rwanda	2 467	9 464	384	2.5	79.8	738	5.3
Saint Helena	39	6	15	0.9	60.0	–	–
Sao Tome and Principe	96	155	161	2.0	41.2	1 522	7.0
Total Central Africa	**529 357**	**116 841**	**22**	**2.9**	**65.2**		
British Indian Ocean Territory	8	1	13	–	–	–	–
Comoros	186	818	440	2.6	62.3	1 144	0.5
Djibouti	2 318	818	35	1.7	13.5	1 966	4.9
Eritrea	10 100	4 692	46	3.7	80.2	682	−1.0
Ethiopia	100 000	81 020	81	2.6	83.7	636	9.0
Kenya	56 914	36 553	64	2.7	79.0	1 467	6.1
Madagascar	58 154	19 159	33	2.8	72.9	878	4.9
Mauritius	203	1 251	616	0.8	57.5	10 571	3.5
Mayotte	37	178	476	–	–	–	–
Réunion	250	796	318	1.4	7.6	–	–
Seychelles	46	86	187	1.2	46.6	15 211	5.3
Somalia	62 734	8 445	13	3.0	64.3	–	–
Uganda	19 710	29 898	152	3.3	87.3	893	5.4
United Republic of Tanzania	88 580	39 458	45	2.5	75.4	995	5.9
Total East Africa	**399 241**	**223 173**	**56**	**2.7**	**79.4**		
Algeria	238 174	33 351	14	1.5	36.1	6 347	3.0
Egypt	99 545	74 166	75	1.8	57.0	4 953	6.8
Libyan Arab Jamahiriya	175 954	6 038	3	2.0	14.9	11 622	5.6
Mauritania	103 070	3 043	3	2.7	59.4	1 890	11.7
Morocco	44 630	30 852	69	1.2	40.7	3 915	8.0
Sudan	237 600	37 707	16	2.2	58.3	1 931	11.8
Tunisia	15 536	10 215	66	1.1	34.3	6 859	5.2
Western Sahara	26 600	461	2	4.8	5.9	–	–
Total Northern Africa	**941 109**	**195 833**	**21**	**1.7**	**48.6**		

Country/area	Land area	Population 2006				GDP 2006	
		Total	Density	Annual growth rate	Rural	Per capita (PPP)	Annual real growth rate
	(1 000 ha)	*(1 000)*	*(population/km²)*	*(%)*	*(% of total)*	*(US$)*	*(%)*
Angola	124 670	16 557	13	2.9	46.0	4 434	18.6
Botswana	56 673	1 858	3	1.3	41.8	12 508	2.1
Lesotho	3 035	1 994	66	0.7	81.0	1 440	7.2
Malawi	9 408	13 570	144	2.6	82.3	700	7.4
Mozambique	78 638	20 971	27	2.1	64.7	739	8.0
Namibia	82 329	2 046	2	1.3	64.3	4 819	2.9
South Africa	121 447	48 282	40	0.7	40.2	9 087	5.0
Swaziland	1 720	1 133	66	0.8	75.6	4 671	2.1
Zambia	74 339	11 696	16	1.9	64.9	1 259	6.2
Zimbabwe	38 685	13 228	34	0.8	63.6	195	−5.4
Total Southern Africa	**590 944**	**131 335**	**22**	**1.5**	**55.1**		
Benin	11 062	8 759	79	3.2	59.5	1 263	4.1
Burkina Faso	27 360	14 358	52	3.1	81.3	1 130	6.4
Cape Verde	403	518	129	2.4	42.0	2 697	6.1
Côte d'Ivoire	31 800	18 914	59	1.8	54.6	1 650	0.9
Gambia	1 000	1 663	166	2.8	45.3	1 130	4.5
Ghana	22 754	23 008	101	2.1	51.5	1 245	6.2
Guinea	24 572	9 181	37	2.0	66.5	1 149	2.8
Guinea-Bissau	2 812	1 645	58	3.1	70.3	478	4.2
Liberia	9 632	3 578	37	4.0	41.2	334	7.8
Mali	122 019	11 968	10	3.1	68.9	1 058	5.3
Niger	126 670	13 736	11	3.6	83.0	629	4.8
Nigeria	91 077	144 719	159	2.4	51.0	1 611	5.2
Senegal	19 253	12 072	63	2.6	58.1	1 585	2.3
Sierra Leone	7 162	5 742	80	2.8	58.6	630	7.4
Togo	5 439	6 410	118	2.8	59.2	776	4.1
Total West Africa	**503 015**	**276 271**	**55**	**2.5**	**56.6**		
Total Africa	**2 963 666**	**943 453**	**32**	**2.3**	**61.2**		
China	932 749	1 328 474	142	0.6	58.7	4 644	10.7
Democratic People's Republic of Korea	12 041	23 707	197	0.4	38.0	–	–
Japan	36 450	127 953	351	0.0	34.0	31 947	2.2
Mongolia	156 650	2 604	2	0.9	43.1	2 887	8.6
Republic of Korea	9 873	48 050	487	0.4	19.0	22 988	5.0
Total East Asia	**1 147 763**	**1 530 788**	**133**	**0.5**	**55.0**		
American Samoa	20	65	325	1.6	8.4	–	–
Australia	768 230	20 530	3	1.1	11.6	35 547	2.5
Cook Islands	24	13	54	−0.5	27.8	–	–

TABLE 1 (CONT.)
Basic data on countries and areas

Country/area	Land area	Population 2006				GDP 2006	
		Total	Density	Annual growth rate	Rural	Per capita (PPP)	Annual real growth rate
	(1 000 ha)	(1 000)	(population/km²)	(%)	(% of total)	(US$)	(%)
Fiji	1 827	833	46	0.6	48.7	4 548	3.6
French Polynesia	366	259	71	1.6	48.3	–	–
Guam	54	171	317	1.8	5.9	–	–
Kiribati	81	93	115	1.1	51.8	3 688	5.8
Marshall Islands	18	57	317	1.8	33.1	6 429	3.0
Micronesia (Federated States of)	70	110	157	0.6	77.6	5 565	–0.7
Nauru	2	10	500	1.6	–	–	–
New Caledonia	1 828	237	13	1.3	35.9	–	–
New Zealand	26 771	4 139	15	1.0	13.7	25 517	1.9
Niue	26	1	4	–0.1	–	–	–
Northern Mariana Islands	46	82	178	2.5	5.3	–	–
Palau	46	20	43	0.5	30.3	14 209	5.7
Papua New Guinea	45 286	6 201	14	2.2	86.5	1 817	2.6
Pitcairn Islands	5	0	1	–	–	–	–
Samoa	283	185	65	1.1	77.4	5 148	2.3
Solomon Islands	2 799	484	17	2.5	82.7	1 839	6.1
Tokelau	1	1	139	0.7	–	–	–
Tonga	72	99	138	0.3	75.7	5 405	1.4
Tuvalu	3	10	333	0.4	40.0	–	–
Vanuatu	1 219	220	18	2.3	76.1	3 768	7.2
Wallis and Futuna Islands	14	15	107	1.2	–	–	–
Total Oceania	**849 091**	**33 835**	**4**	**1.3**	**29.3**		
Bangladesh	13 017	155 990	1 198	1.8	74.5	1 155	6.6
Bhutan	4 700	648	14	1.7	88.6	4 010	8.5
India	297 319	1 151 751	387	1.5	71.0	2 469	9.2
Maldives	30	300	1 000	1.7	69.9	5 008	23.5
Nepal	14 300	27 641	193	2.0	83.7	999	2.8
Pakistan	77 088	160 943	209	1.8	64.7	2 361	6.9
Sri Lanka	6 463	19 207	297	0.5	84.9	3 747	7.4
Total South Asia	**412 917**	**1 516 480**	**367**	**1.6**	**71.1**		
Brunei Darussalam	527	381	72	2.1	26.1	49 898	5.1
Cambodia	17 652	14 196	80	1.7	79.7	1 619	10.8
Indonesia	181 157	228 864	126	1.2	50.8	3 454	5.5
Lao People's Democratic Republic	23 080	5 759	25	1.7	79.0	1 980	7.6
Malaysia	32 855	26 113	79	1.8	31.8	12 536	5.9
Myanmar	65 755	48 379	74	0.9	68.7	979	4.1
Philippines	29 817	86 263	289	2.0	36.6	3 153	5.4

Country/area	Land area	Population 2006				GDP 2006	
		Total	Density	Annual growth rate	Rural	Per capita (PPP)	Annual real growth rate
	(1 000 ha)	(1 000)	(population/km²)	(%)	(% of total)	(US$)	(%)
Singapore	69	4 381	6 358	1.2	0.0	44 708	7.9
Thailand	51 089	63 443	124	0.7	67.4	7 599	5.0
Timor-Leste	1 487	1 113	75	4.3	73.1	2 141	−1.6
Viet Nam	31 007	86 205	278	1.4	73.1	2 363	8.2
Total Southeast Asia	**434 495**	**565 097**	**130**	**1.3**	**55.2**		
Total Asia and the Pacific	**2 844 265**	**3 646 200**	**128**	**1.1**	**61.5**		
Belarus	20 748	9 742	47	−0.5	27.3	9 732	9.9
Republic of Moldova	3 287	3 832	117	−1.1	53.0	2 377	4.0
Russian Federation	1 638 139	143 221	9	−0.5	27.1	13 116	6.7
Ukraine	57 938	46 557	80	−0.8	32.0	6 212	7.1
Total CIS countries	**1 720 112**	**203 352**	**12**	**−0.6**	**28.7**		
Albania	2 740	3 172	116	0.6	53.9	5 886	5.0
Bosnia and Herzegovina	5 120	3 926	77	0.3	53.7	6 488	6.0
Bulgaria	10 864	7 692	71	−0.7	29.7	10 274	6.1
Croatia	5 592	4 556	81	0.1	43.2	14 309	4.8
Czech Republic	7 726	10 188	132	0.0	26.5	22 118	6.1
Estonia	4 239	1 339	32	−0.4	30.9	18 969	11.4
Hungary	8 961	10 058	112	−0.3	33.3	18 277	3.9
Latvia	6 229	2 289	37	−0.5	32.1	15 350	11.9
Lithuania	6 268	3 408	54	−0.5	33.4	15 738	7.7
Montenegro	1 380	608	44	−0.3	–	9 034	16.2
Poland	30 633	38 140	125	−0.1	37.8	14 836	6.1
Romania	22 998	21 531	94	−0.4	46.1	10 431	7.7
Serbia	8 820	9 875	112	0.1	–	9 434	5.7
Slovakia	4 810	5 388	112	0.0	43.7	17 730	8.3
Slovenia	2 014	2 000	99	0.1	48.8	24 356	5.2
The former Yugoslav Republic of Macedonia	2 543	2 036	80	0.1	30.4	7 850	3.0
Total Eastern Europe	**130 937**	**126 206**	**96**	**−0.2**	**39.4**		
Andorra	47	74	157	1.4	9.7	–	–
Austria	8 245	8 327	101	0.4	33.9	36 049	3.1
Belgium	3 023	10 430	345	0.3	2.8	33 543	3.2
Channel Islands	19	148	779	0.4	69.4	–	–
Denmark	4 243	5 430	128	0.3	14.3	35 692	3.2
Faeroe Islands	140	48	34	0.6	61.0	–	–
Finland	30 459	5 261	17	0.3	38.8	33 022	5.5
France	55 010	61 329	111	0.6	23.1	31 992	2.0
Germany	34 877	82 640	237	0.0	24.7	32 322	2.8

TABLE 1 (CONT.)

Basic data on countries and areas

Country/area	Land area	Population 2006				GDP 2006	
		Total	Density	Annual growth rate	Rural	Per capita (PPP)	Annual real growth rate
	(1 000 ha)	*(1 000)*	*(population/km²)*	*(%)*	*(% of total)*	*(US$)*	*(%)*
Gibraltar	1	29	2 900	0.1	0.0	–	–
Greece	12 890	11 122	86	0.2	40.9	27 333	4.3
Holy See	–	1	–	–0.1	0.0	–	–
Iceland	10 025	298	3	1.0	7.1	36 923	2.6
Ireland	6 889	4 221	61	1.9	39.2	40 268	5.7
Isle of Man	57	78	137	–0.2	48.1	–	–
Italy	29 411	58 778	200	0.2	32.2	29 053	1.9
Liechtenstein	16	34	213	0.9	85.4	–	–
Luxembourg	259	461	178	1.1	17.3	75 611	6.2
Malta	32	404	1 263	0.5	4.5	21 720	3.4
Monaco	2	32	1 600	1.1	0.0	–	–
Netherlands	3 388	16 378	483	0.3	19.3	36 560	2.9
Norway	30 428	4 668	15	0.6	22.5	50 078	2.9
Portugal	9 150	10 578	116	0.5	41.8	20 784	1.3
San Marino	6	30	500	0.8	2.5	–	–
Spain	49 919	43 886	88	1.1	23.2	28 649	3.9
Sweden	41 033	9 078	22	0.4	15.7	34 193	4.2
Switzerland	4 000	7 454	186	0.4	24.4	37 194	3.2
United Kingdom	24 193	60 512	250	0.1	10.2	33 087	2.8
Total Western Europe	**357 762**	**401 729**	**112**	**0.4**	**23.4**		
Total Europe	**2 208 811**	**731 287**	**33**	**0.0**	**27.7**		
Anguilla	9	12	133	1.5	–	–	–
Antigua and Barbuda	44	84	191	1.2	60.4	16 578	11.5
Aruba	18	103	572	1.0	53.3	–	–
Bahamas	1 001	327	33	1.2	9.4	23 927	3.4
Barbados	43	292	679	0.3	46.7	18 145	3.9
Bermuda	5	64	1 280	0.3	0.0	–	–
British Virgin Islands	15	22	147	1.2	36.4	–	–
Cayman Islands	26	46	177	2.2	0.0	–	–
Cuba	10 982	11 266	103	0.1	24.6	–	–
Dominica	75	67	89	0.8	26.8	9 236	4.0
Dominican Republic	4 838	9 614	199	1.5	32.5	5 866	10.7
Grenada	34	105	309	1.0	69.3	9 415	0.7
Guadeloupe	169	441	261	0.7	0.2	–	–
Haiti	2 756	9 445	343	1.6	60.5	1 224	2.3
Jamaica	1 083	2 698	249	0.6	46.6	7 567	2.5
Martinique	106	397	375	0.5	3.8	–	–
Montserrat	10	5	50	3.5	–	–	–

Country/area	Land area	Population 2006				GDP 2006	
		Total	Density	Annual growth rate	Rural	Per capita (PPP)	Annual real growth rate
	(1 000 ha)	(1 000)	(population/km²)	(%)	(% of total)	(US$)	(%)
Netherlands Antilles	80	188	235	1.1	29.3	–	–
Puerto Rico	887	3 968	447	0.6	2.2	–	–
Saint Kitts and Nevis	26	49	188	1.1	67.8	14 886	5.8
Saint Lucia	61	163	267	1.2	72.3	9 992	4.5
Saint Vincent and the Grenadines	39	119	305	0.5	53.7	8 916	6.9
Trinidad and Tobago	513	1 328	259	0.4	87.5	17 717	12.0
Turks and Caicos Islands	43	25	58	4.2	53.8	–	–
United States Virgin Islands	35	111	317	0.0	5.6	–	–
Total Caribbean	**22 898**	**40 939**	**179**	**0.9**	**36.1**		
Belize	2 281	281	12	2.2	51.5	7 846	5.6
Costa Rica	5 106	4 398	86	1.6	37.8	9 564	8.2
El Salvador	2 072	6 762	326	1.4	39.9	5 765	4.2
Guatemala	10 843	13 028	120	2.5	52.3	5 175	4.5
Honduras	11 189	6 968	62	2.0	53.0	3 543	6.0
Nicaragua	12 140	5 532	46	1.3	40.6	2 789	3.7
Panama	7 443	3 287	44	1.7	28.4	9 255	8.1
Total Central America	**51 074**	**40 256**	**79**	**1.9**	**45.2**		
Argentina	273 669	39 134	14	1.0	9.7	11 985	8.5
Bolivia	108 438	9 353	9	1.9	35.3	3 937	4.6
Brazil	845 942	189 322	22	1.3	15.3	8 949	3.7
Chile	74 880	16 465	22	1.0	12.1	13 030	4.0
Colombia	110 950	45 558	41	1.4	27.0	6 378	6.8
Ecuador	27 684	13 201	48	1.1	36.7	7 145	3.9
Falkland Islands	1 217	2	0	0.4	–	–	–
French Guiana	8 815	197	2	2.6	24.6	–	–
Guyana	19 685	739	4	0.1	71.7	3 547	4.8
Paraguay	39 730	6 015	15	1.9	40.9	4 034	4.3
Peru	128 000	27 588	22	1.2	27.2	7 092	7.7
South Georgia and the South Sandwich Islands	409	0	0	–	–	–	–
Suriname	15 600	455	3	0.7	25.8	7 984	5.8
Uruguay	17 502	3 331	19	0.2	7.9	10 203	7.0
Venezuela (Bolivarian Republic of)	88 205	27 191	31	1.7	6.3	11 060	10.3
Total South America	**1 760 726**	**378 551**	**21**	**1.3**	**17.9**		
Total Latin America and the Caribbean	**1 834 698**	**459 746**	**25**	**1.3**	**21.9**		
Canada	909 351	32 576	4	0.9	19.8	36 713	2.8
Greenland	41 045	57	0	0.3	16.8	–	–
Mexico	194 395	105 342	54	1.0	23.7	12 177	4.8

TABLE 1 (CONT.)

Basic data on countries and areas

Country/area	Land area	Population 2006				GDP 2006	
		Total	Density	Annual growth rate	Rural	Per capita (PPP)	Annual real growth rate
	(1 000 ha)	*(1 000)*	*(population/km²)*	*(%)*	*(% of total)*	*(US$)*	*(%)*
Saint Pierre and Miquelon	23	6	26	0.8	16.7	–	–
United States of America	916 192	302 841	33	1.0	18.9	43 968	2.9
Total North America	**2 061 006**	**440 822**	**21**	**1.0**	**20.1**		
Armenia	2 820	3 009	107	−0.3	36.0	4 879	13.3
Azerbaijan	8 266	8 406	102	0.6	48.4	6 280	30.6
Georgia	6 949	4 432	64	−0.9	47.7	4 010	9.4
Kazakhstan	269 970	15 314	6	0.7	42.4	9 832	10.7
Kyrgyzstan	19 180	5 258	27	1.1	64.0	1 813	2.7
Tajikistan	13 996	6 639	47	1.4	75.4	1 610	7.0
Turkmenistan	46 993	4 899	10	1.4	53.4	4 570	11.1
Uzbekistan	42 540	26 980	63	1.5	63.3	2 192	7.3
Total Central Asia	**410 714**	**74 937**	**18**	**1.0**	**55.8**		
Afghanistan	65 209	26 087	40	4.1	76.7	917	5.3
Bahrain	71	738	1 039	1.9	3.3	33 451	6.5
Cyprus	924	845	91	1.1	30.5	25 882	4.0
Iran (Islamic Republic of)	162 855	70 270	43	1.2	32.6	9 906	4.6
Iraq	43 737	28 505	65	1.8	33.2	–	–
Israel	2 164	6 809	315	1.7	8.4	24 097	5.1
Jordan	8 824	5 728	65	3.3	17.4	4 628	5.7
Kuwait	1 782	2 778	156	0.7	1.7	43 551	6.3
Lebanon	1 023	4 055	396	1.1	13.3	9 741	0.0
Occupied Palestinian Territory	602	3 889	646	3.4	28.3	3 605	1.4
Oman	30 950	2 546	8	1.6	28.5	22 152	6.8
Qatar	1 100	821	75	3.1	4.5	70 772	10.3
Saudi Arabia	214 969	24 174	11	2.4	18.8	22 296	4.3
Syrian Arab Republic	18 378	19 407	106	2.7	49.2	4 225	5.1
Turkey	76 963	73 921	96	1.3	32.2	8 417	6.1
United Arab Emirates	8 360	4 248	51	3.5	23.3	35 882	9.4
Yemen	52 797	21 732	41	3.0	72.3	2 264	3.3
Total Western Asia	**690 708**	**296 553**	**43**	**2.0**	**37.5**		
Total Western and Central Asia	**1 101 422**	**371 490**	**34**	**1.8**	**41.2**		
TOTAL WORLD	**13 013 868**	**6 592 998**	**51**	**1.2**	**51.0**		

SOURCES: FAOSTAT (ResourceSTAT and PopSTAT), World Bank (World Development Indicators) and IMF (World Economic Outlook database), last accessed 28 August 2008.

TABLE 2
Forest area and area change

Country/area	Extent of forest, 2005			Annual change rate			
	Forest area	% of land area	Area per 1 000 people	1990–2000		2000–2005	
	(1 000 ha)	(%)	(ha)	(1 000 ha)	(%)	(1 000 ha)	(%)
Burundi	152	5.9	19	−9	−3.7	−9	−5.2
Cameroon	21 245	45.6	1 169	−220	−0.9	−220	−1.0
Central African Republic	22 755	36.5	5 337	−30	−0.1	−30	−0.1
Chad	11 921	9.5	1 139	−79	−0.6	−79	−0.7
Congo	22 471	65.8	6 091	−17	−0.1	−17	−0.1
Democratic Republic of the Congo	133 610	58.9	2 203	−532	−0.4	−319	−0.2
Equatorial Guinea	1 632	58.2	3 297	−15	−0.8	−15	−0.9
Gabon	21 775	84.5	16 622	−10	0.0	−10	0.0
Rwanda	480	19.5	51	3	0.8	27	6.9
Saint Helena	2	6.5	333	0	0.0	0	0.0
Sao Tome and Principe	27	28.4	177	0	0.0	0	0.0
Total Central Africa	**236 070**	**44.6**	**2 020**	**−910**	**−0.37**	**−673**	**−0.28**
British Indian Ocean Territory	3	32.5	2 600	0	0.0	0	0.0
Comoros	5	2.9	7	0	−4.0	−1	−7.4
Djibouti	6	0.2	7	0	0.0	0	0.0
Eritrea	1 554	15.4	331	−5	−0.3	−4	−0.3
Ethiopia	13 000	11.9	160	−141	−1.0	−141	−1.1
Kenya	3 522	6.2	96	−13	−0.3	−12	−0.3
Madagascar	12 838	22.1	670	−67	−0.5	−37	−0.3
Mauritius	37	18.2	30	0	−0.3	0	−0.5
Mayotte	5	14.7	31	0	−0.4	0	−0.4
Réunion	84	33.6	106	0	−0.1	−1	−0.7
Seychelles	40	88.9	465	0	0.0	0	0.0
Somalia	7 131	11.4	844	−77	−1.0	−77	−1.0
Uganda	3 627	18.4	121	−87	−1.9	−86	−2.2
United Republic of Tanzania	35 257	39.9	894	−412	−1.0	−412	−1.1
Total East Africa	**77 109**	**18.9**	**346**	**−801**	**−0.94**	**−771**	**−0.97**
Algeria	2 277	1.0	68	35	1.8	27	1.2
Egypt	67	0.1	1	2	3.0	2	2.6
Libyan Arab Jamahiriya	217	0.1	36	0	0.0	0	0.0
Mauritania	267	0.3	88	−10	−2.7	−10	−3.4
Morocco	4 364	9.8	141	4	0.1	7	0.2
Sudan	67 546	28.4	1 791	−589	−0.8	−589	−0.8
Tunisia	1 056	6.8	103	32	4.1	19	1.9
Western Sahara	1 011	3.8	2 193	0	0.0	0	0.0
Total Northern Africa	**76 805**	**8.2**	**392**	**−526**	**−0.64**	**−544**	**−0.69**

TABLE 2 (CONT.)
Forest area and area change

Country/area	Extent of forest, 2005			Annual change rate			
	Forest area	% of land area	Area per 1 000 people	1990–2000		2000–2005	
	(1 000 ha)	*(%)*	*(ha)*	*(1 000 ha)*	*(%)*	*(1 000 ha)*	*(%)*
Angola	59 104	47.4	3 570	−125	−0.2	−125	−0.2
Botswana	11 943	21.1	6 428	−118	−0.9	−118	−1.0
Lesotho	8	0.3	4	0	3.4	0	2.7
Malawi	3 402	36.2	251	−33	−0.9	−33	−0.9
Mozambique	19 262	24.6	919	−50	−0.3	−50	−0.3
Namibia	7 661	9.3	3 744	−73	−0.9	−74	−0.9
South Africa	9 203	7.6	191	0	0.0	0	0.0
Swaziland	541	31.5	477	5	0.9	5	0.9
Zambia	42 452	57.1	3 630	−445	−0.9	−445	−1.0
Zimbabwe	17 540	45.3	1 326	−313	−1.5	−313	−1.7
Total Southern Africa	**171 116**	**29.0**	**1 303**	**−1 152**	**−0.63**	**−1 154**	**−0.66**
Benin	2 351	21.3	268	−65	−2.1	−65	−2.5
Burkina Faso	6 794	29.0	473	−24	−0.3	−24	−0.3
Cape Verde	84	20.7	161	2	3.6	0	0.4
Côte d'Ivoire	10 405	32.7	550	11	0.1	15	0.1
Gambia	471	41.7	283	2	0.4	2	0.4
Ghana	5 517	24.2	240	−135	−2.0	−115	−2.0
Guinea	6 724	27.4	732	−50	−0.7	−36	−0.5
Guinea-Bissau	2 072	73.7	1 259	−10	−0.4	−10	−0.5
Liberia	3 154	32.7	881	−60	−1.6	−60	−1.8
Mali	12 572	10.3	1 050	−100	−0.7	−100	−0.8
Niger	1 266	1.0	92	−62	−3.7	−12	−1.0
Nigeria	11 089	12.2	77	−410	−2.7	−410	−3.3
Senegal	8 673	45.0	718	−45	−0.5	−45	−0.5
Sierra Leone	2 754	38.5	480	−19	−0.7	−19	−0.7
Togo	386	7.1	60	−20	−3.4	−20	−4.5
Total West Africa	**74 312**	**14.9**	**269**	**−985**	**−1.17**	**−899**	**−1.17**
Total Africa	**635 412**	**21.4**	**673**	**−4 375**	**−0.64**	**−4 040**	**−0.62**
China	197 290	21.2	149	1 986	1.2	4 058	2.2
Democratic People's Republic of Korea	6 187	51.4	261	−138	−1.8	−127	−1.9
Japan	24 868	68.2	194	−7	0.0	−2	0.0
Mongolia	10 252	6.5	3 937	−83	−0.7	−83	−0.8
Republic of Korea	6 265	63.5	130	−7	−0.1	−7	−0.1
Total East Asia	**244 862**	**21.3**	**160**	**1 751**	**0.81**	**3 840**	**1.65**

Country/area	Extent of forest, 2005			Annual change rate			
	Forest area	% of land area	Area per 1 000 people	1990–2000		2000–2005	
	(1 000 ha)	(%)	(ha)	(1 000 ha)	(%)	(1 000 ha)	(%)
American Samoa	18	89.4	275	0	−0.2	0	−0.2
Australia	163 678	21.3	7 973	−326	−0.2	−193	−0.1
Cook Islands	16	66.5	1 192	0	0.4	0	0.0
Fiji	1 000	54.7	1 200	2	0.2	0	0.0
French Polynesia	105	28.7	405	0	0.0	0	0.0
Guam	26	47.1	151	0	0.0	0	0.0
Kiribati	2	3.0	24	0	0.0	0	0.0
Marshall Islands	–	–	–	–	–	–	–
Micronesia (Federated States of)	63	90.6	576	0	0.0	0	0.0
Nauru	0	0.0	0	0	0.0	0	0.0
New Caledonia	717	39.2	3 025	0	0.0	0	0.0
New Zealand	8 309	31.0	2 007	51	0.6	17	0.2
Niue	14	54.2	14 100	0	−1.3	0	−1.4
Northern Mariana Islands	33	72.4	406	0	−0.3	0	−0.3
Palau	40	87.6	2 015	0	0.4	0	0.4
Papua New Guinea	29 437	65.0	4 747	−139	−0.5	−139	−0.5
Pitcairn Islands	4	83.3	52 239	0	0.0	0	0.0
Samoa	171	60.4	924	4	2.8	0	0.0
Solomon Islands	2 172	77.6	4 488	−40	−1.5	−40	−1.7
Tokelau	0	0.0	0	0	0.0	0	0.0
Tonga	4	5.0	36	0	0.0	0	0.0
Tuvalu	1	33.3	100	0	0.0	0	0.0
Vanuatu	440	36.1	1 998	0	0.0	0	0.0
Wallis and Futuna Islands	5	35.3	328	0	−0.8	0	−2.0
Total Oceania	**206 254**	**24.3**	**6 096**	**−448**	**−0.21**	**−356**	**−0.17**
Bangladesh	871	6.7	6	0	0.0	−2	−0.3
Bhutan	3 195	68.0	4 931	11	0.3	11	0.3
India	67 701	22.8	59	362	0.6	29	0.0
Maldives	1	3.0	3	0	0.0	0	0.0
Nepal	3 636	25.4	132	−92	−2.1	−53	−1.4
Pakistan	1 902	2.5	12	−41	−1.8	−43	−2.1
Sri Lanka	1 933	29.9	101	−27	−1.2	−30	−1.5
Total South Asia	**79 239**	**19.2**	**52**	**213**	**0.27**	**−88**	**−0.11**
Brunei Darussalam	278	52.8	730	−3	−0.8	−2	−0.7
Cambodia	10 447	59.2	736	−141	−1.1	−219	−2.0
Indonesia	88 495	48.8	387	−1 872	−1.7	−1 871	−2.0
Lao People's Democratic Republic	16 142	69.9	2 803	−78	−0.5	−78	−0.5
Malaysia	20 890	63.6	800	−79	−0.4	−140	−0.7

TABLE 2 (CONT.)
Forest area and area change

Country/area	Extent of forest, 2005			Annual change rate			
	Forest area	% of land area	Area per 1 000 people	1990–2000		2000–2005	
	(1 000 ha)	(%)	(ha)	(1 000 ha)	(%)	(1 000 ha)	(%)
Myanmar	32 222	49.0	666	−467	−1.3	−466	−1.4
Philippines	7 162	24.0	83	−263	−2.8	−157	−2.1
Singapore	2	3.4	1	0	0.0	0	0.0
Thailand	14 520	28.4	229	−115	−0.7	−59	−0.4
Timor-Leste	798	53.7	717	−11	−1.2	−11	−1.3
Viet Nam	12 931	39.7	150	236	2.3	241	2.0
Total Southeast Asia	**203 887**	**46.8**	**361**	**−2 790**	**−1.20**	**−2 763**	**−1.30**
Total Asia and the Pacific	**734 243**	**25.8**	**201**	**−1 275**	**−0.17**	**633**	**0.09**
Belarus	7 894	38.0	810	47	0.6	9	0.1
Republic of Moldova	329	10.0	86	1	0.2	1	0.2
Russian Federation	808 790	47.9	5 647	32	0.0	−96	0.0
Ukraine	9 575	16.5	206	24	0.3	13	0.1
Total CIS countries	**826 588**	**46.7**	**4 065**	**103**	**0.01**	**−73**	**−0.01**
Albania	794	29.0	250	−2	−0.3	5	0.6
Bosnia and Herzegovina	2 185	43.1	557	−3	−0.1	0	0.0
Bulgaria	3 625	32.8	471	5	0.1	50	1.4
Croatia	2 135	38.2	469	1	0.1	1	0.1
Czech Republic	2 648	34.3	260	1	0.0	2	0.1
Estonia	2 284	53.9	1 706	8	0.4	8	0.4
Hungary	1 976	21.5	196	11	0.6	14	0.7
Latvia	2 941	47.4	1 285	11	0.4	11	0.4
Lithuania	2 099	33.5	616	8	0.4	16	0.8
Poland	9 192	30.0	241	18	0.2	27	0.3
Romania	6 370	27.7	296	−1	0.0	1	0.0
Serbia and Montenegro	2 694	26.4	256	9	0.3	9	0.3
Slovakia	1 929	40.1	358	0	0.0	2	0.1
Slovenia	1 264	62.8	632	5	0.4	5	0.4
The former Yugoslav Republic of Macedonia	906	35.8	445	0	0.0	0	0.0
Total Eastern Europe	**43 042**	**32.8**	**341**	**71**	**0.17**	**150**	**0.35**
Andorra	16	35.6	216	0	0.0	0	0.0
Austria	3 862	46.7	464	6	0.2	5	0.1
Belgium	667	22.0	64	−1	−0.1	0	0.0
Channel Islands	1	4.1	5	0	0.0	0	0.0
Denmark	500	11.8	92	4	0.9	3	0.6
Faeroe Islands	0	0.1	2	0	0.0	0	0.0

Country/area	Extent of forest, 2005			Annual change rate			
	Forest area	% of land area	Area per 1 000 people	1990–2000		2000–2005	
	(1 000 ha)	(%)	(ha)	(1 000 ha)	(%)	(1 000 ha)	(%)
Finland	22 500	73.9	4 277	28	0.1	5	0.0
France	15 554	28.3	254	81	0.5	41	0.3
Germany	11 076	31.7	134	34	0.3	0	0.0
Gibraltar	0	0.0	0	0	0.0	0	0.0
Greece	3 752	29.1	337	30	0.9	30	0.8
Holy See	0	0.0	0	0	0.0	0	0.0
Iceland	46	0.0	154	1	4.3	2	3.9
Ireland	669	9.7	158	17	3.3	12	1.9
Isle of Man	3	6.1	44	0	0.0	0	0.0
Italy	9 979	33.9	170	106	1.2	106	1.1
Liechtenstein	7	43.1	203	0	0.6	0	0.0
Luxembourg	87	33.5	188	0	0.1	0	0.0
Malta	0	1.1	1	0	0.0	0	0.0
Monaco	0	0.0	0	0	0.0	0	0.0
Netherlands	365	10.8	22	2	0.4	1	0.3
Norway	9 387	30.7	2 011	17	0.2	17	0.2
Portugal	3 783	41.3	358	48	1.5	40	1.1
San Marino	0	1.6	3	0	0.0	0	0.0
Spain	17 915	35.9	408	296	2.0	296	1.7
Sweden	27 528	66.9	3 032	11	0.0	11	0.0
Switzerland	1 221	30.9	164	4	0.4	4	0.4
United Kingdom	2 845	11.8	47	18	0.7	10	0.4
Total Western Europe	**131 763**	**36.8**	**328**	**703**	**0.56**	**583**	**0.45**
Total Europe	**1 001 394**	**44.3**	**1 369**	**877**	**0.09**	**661**	**0.07**
Anguilla	6	71.4	458	0	0.0	0	0.0
Antigua and Barbuda	9	21.4	112	0	0.0	0	0.0
Aruba	0	2.2	4	0	0.0	0	0.0
Bahamas	515	51.5	1 575	0	0.0	0	0.0
Barbados	2	4.0	6	0	0.0	0	0.0
Bermuda	1	20.0	16	0	0.0	0	0.0
British Virgin Islands	4	24.4	166	0	−0.1	0	−0.1
Cayman Islands	12	48.4	270	0	0.0	0	0.0
Cuba	2 713	24.7	241	38	1.7	56	2.2
Dominica	46	61.3	686	0	−0.5	0	−0.6
Dominican Republic	1 376	28.4	143	0	0.0	0	0.0
Grenada	4	12.2	39	0	0.0	0	0.0
Guadeloupe	80	47.2	181	0	−0.3	0	−0.3
Haiti	105	3.8	11	−1	−0.6	−1	−0.7
Jamaica	339	31.3	126	0	−0.1	0	−0.1

TABLE 2 (CONT.)
Forest area and area change

Country/area	Extent of forest, 2005			Annual change rate			
	Forest area	% of land area	Area per 1 000 people	1990–2000		2000–2005	
	(1 000 ha)	(%)	(ha)	(1 000 ha)	(%)	(1 000 ha)	(%)
Martinique	47	43.9	117	0	0.0	0	0.0
Montserrat	4	35.0	700	0	0.0	0	0.0
Netherlands Antilles	1	1.5	6	0	0.0	0	0.0
Puerto Rico	408	46.0	103	0	0.1	0	0.0
Saint Kitts and Nevis	5	14.7	108	0	0.0	0	0.0
Saint Lucia	17	27.9	104	0	0.0	0	0.0
Saint Vincent and the Grenadines	11	27.4	90	0	0.8	0	0.8
Trinidad and Tobago	226	44.1	170	−1	−0.3	0	−0.2
Turks and Caicos Islands	34	80.0	1 376	0	0.0	0	0.0
United States Virgin Islands	10	27.9	86	0	−1.3	0	−1.8
Total Caribbean	**5 974**	**26.1**	**146**	**36**	**0.65**	**54**	**0.92**
Belize	1 653	72.5	5 883	0	0.0	0	0.0
Costa Rica	2 391	46.8	544	−19	−0.8	3	0.1
El Salvador	298	14.4	44	−5	−1.5	−5	−1.7
Guatemala	3 938	36.3	302	−54	−1.2	−54	−1.3
Honduras	4 648	41.5	667	−196	−3.0	−156	−3.1
Nicaragua	5 189	42.7	938	−100	−1.6	−70	−1.3
Panama	4 294	57.7	1 306	−7	−0.2	−3	−0.1
Total Central America	**22 411**	**43.9**	**557**	**−380**	**−1.47**	**−285**	**−1.23**
Argentina	33 021	12.1	844	−149	−0.4	−150	−0.4
Bolivia	58 740	54.2	6 280	−270	−0.4	−270	−0.5
Brazil	477 698	57.2	2 523	−2 681	−0.5	−3 103	−0.6
Chile	16 121	21.5	979	57	0.4	57	0.4
Colombia	60 728	58.5	1 333	−48	−0.1	−47	−0.1
Ecuador	10 853	39.2	822	−198	−1.5	−198	−1.7
Falkland Islands	0	0.0	0	0	0.0	0	0.0
French Guiana	8 063	91.8	40 929	−3	0.0	0	0.0
Guyana	15 104	76.7	20 438	0	0.0	0	0.0
Paraguay	18 475	46.5	3 071	−179	−0.9	−179	−0.9
Peru	68 742	53.7	2 492	−94	−0.1	−94	−0.1
South Georgia and the South Sandwich Islands	0	0.0	0	0	0.0	0	0.0
Suriname	14 776	94.7	32 475	0	0.0	0	0.0
Uruguay	1 506	8.6	452	50	4.5	19	1.3
Venezuela (Bolivarian Republic of)	47 713	54.1	1 755	−288	−0.6	−288	−0.6
Total South America	**831 540**	**47.7**	**2 197**	**−3 802**	**−0.44**	**−4 251**	**−0.50**
Total Latin America and the Caribbean	**859 925**	**47.3**	**1 870**	**−4 147**	**−0.46**	**−4 483**	**−0.51**

Country/area	Extent of forest, 2005			Annual change rate			
	Forest area	% of land area	Area per 1 000 people	1990–2000		2000–2005	
	(1 000 ha)	(%)	(ha)	(1 000 ha)	(%)	(1 000 ha)	(%)
Canada	310 134	33.6	9 520	0	0.0	0	0.0
Greenland	0	0.0	4	0	0.0	0	0.0
Mexico	64 238	33.7	610	−348	−0.5	−260	−0.4
Saint Pierre and Miquelon	3	13.0	500	0	0.0	0	0.0
United States of America	303 089	33.1	1 001	365	0.1	159	0.1
Total North America	**677 464**	**32.7**	**1 537**	**17**	**0.00**	**−101**	**−0.01**
Armenia	283	10.0	94	−4	−1.3	−4	−1.5
Azerbaijan	936	11.3	111	0	0.0	0	0.0
Georgia	2 760	39.7	623	0	0.0	0	0.0
Kazakhstan	3 337	1.2	218	−6	−0.2	−6	−0.2
Kyrgyzstan	869	4.5	165	2	0.3	2	0.3
Tajikistan	410	2.9	62	0	0.0	0	0.0
Turkmenistan	4 127	8.8	842	0	0.0	0	0.0
Uzbekistan	3 295	8.0	122	17	0.5	17	0.5
Total Central Asia	**16 017**	**3.9**	**214**	**9**	**0.06**	**9**	**0.06**
Afghanistan	867	1.3	33	−29	−2.5	−30	−3.1
Bahrain	0	0.6	1	0	5.6	0	3.8
Cyprus	174	18.9	206	1	0.7	0	0.2
Iran (Islamic Republic of)	11 075	6.8	158	0	0.0	0	0.0
Iraq	822	1.9	29	1	0.2	1	0.1
Israel	171	8.3	25	1	0.6	1	0.8
Jordan	83	0.9	14	0	0.0	0	0.0
Kuwait	6	0.3	2	0	3.5	0	2.7
Lebanon	137	13.3	34	1	0.8	1	0.8
Occupied Palestinian Territory	9	1.5	2	0	0.0	0	0.0
Oman	2	0.0	1	0	0.0	0	0.0
Qatar	0	0.0	0	0	0.0	0	0.0
Saudi Arabia	2 728	1.3	113	0	0.0	0	0.0
Syrian Arab Republic	461	2.5	24	6	1.5	6	1.3
Turkey	10 175	13.2	138	37	0.4	25	0.2
United Arab Emirates	312	3.7	73	7	2.4	0	0.1
Yemen	549	1.0	25	0	0.0	0	0.0
Total Western Asia	**27 571**	**4.0**	**93**	**25**	**0.09**	**5**	**0.02**
Total Western and Central Asia	**43 588**	**4.0**	**117**	**34**	**0.08**	**14**	**0.03**
TOTAL WORLD	**3 952 025**	**30.3**	**599**	**−8 868**	**−0.22**	**−7 317**	**−0.18**

SOURCE: FAO, 2006a.

TABLE 3
Forest growing stock, biomass and carbon

Country/area	Growing stock			Biomass		Carbon in biomass	
	Per hectare	Total	Commercial	Per hectare	Total	Per hectare	Total
	(m³/ha)	*(million m³)*	*(%)*	*(tonnes/ha)*	*(million tonnes)*	*(tonnes/ha)*	*(million tonnes)*
Burundi	–	–	–	–	–	–	–
Cameroon	62	1 313	10	179	3 804	90	1 902
Central African Republic	167	3 801	–	246	5 604	123	2 801
Chad	18	218	38	40	471	20	236
Congo	203	4 551	30	461	10 361	231	5 181
Democratic Republic of the Congo	231	30 833	–	347	46 346	173	23 173
Equatorial Guinea	66	107	–	142	231	70	115
Gabon	223	4 845	–	335	7 285	167	3 643
Rwanda	183	88	95	183	88	92	44
Saint Helena	–	–	–	–	–	–	–
Sao Tome and Principe	164	5	100	336	9	168	5
Total Central Africa	**194**	**45 760**		**315**	**74 199**	**157**	**37 100**
British Indian Ocean Territory	–	–	–	–	–	–	–
Comoros	247	1	27	284	2	141	1
Djibouti	32	0	–	156	1	78	0
Eritrea	–	–	–	–	–	–	–
Ethiopia	22	285	25	39	503	19	252
Kenya	80	281	11	190	669	95	335
Madagascar	171	2 201	28	488	6 259	244	3 130
Mauritius	82	3	68	212	8	106	4
Mayotte	–	–	–	–	–	–	–
Réunion	–	–	–	–	–	–	–
Seychelles	75	3	12	175	7	93	4
Somalia	22	157	0	108	773	54	387
Uganda	43	156	15	76	276	38	138
United Republic of Tanzania	36	1 264	73	128	4 509	64	2 254
Total East Africa	**58**	**4 351**		**172**	**13 007**	**86**	**6 504**
Algeria	76	174	22	100	227	50	114
Egypt	120	8	–	212	14	106	7
Libyan Arab Jamahiriya	36	8	–	59	13	30	6
Mauritania	20	5	–	50	13	25	7
Morocco	44	191	100	110	480	55	240
Sudan	14	939	–	45	3 061	23	1 531
Tunisia	26	27	2	19	20	9	10
Western Sahara	38	38	–	50	50	25	25
Total Northern Africa	**18**	**1 390**		**51**	**3 879**	**25**	**1 940**

Country/area	Growing stock			Biomass		Carbon in biomass	
	Per hectare	Total	Commercial	Per hectare	Total	Per hectare	Total
	(m³/ha)	(million m³)	(%)	(tonnes/ha)	(million tonnes)	(tonnes/ha)	(million tonnes)
Angola	39	2 291	1	163	9 659	82	4 829
Botswana	16	197	–	24	283	12	142
Lesotho	–	–	–	–	–	–	–
Malawi	110	373	–	95	322	47	161
Mozambique	26	496	14	63	1 213	31	606
Namibia	24	184	–	60	462	30	231
South Africa	69	635	38	179	1 648	90	824
Swaziland	35	19	–	87	47	43	23
Zambia	31	1 307	7	54	2 312	27	1 156
Zimbabwe	34	600	4	61	1 069	31	535
Total Southern Africa	**36**	**6 102**		**99**	**17 014**	**50**	**8 507**
Benin	–	–	–	–	–	–	–
Burkina Faso	35	238	5	88	596	44	298
Cape Verde	144	12	80	189	16	95	8
Côte d'Ivoire	258	2 683	20	386	4 014	179	1 864
Gambia	37	18	–	141	66	70	33
Ghana	58	321	53	180	993	90	496
Guinea	77	520	–	189	1 272	95	636
Guinea-Bissau	24	50	20	59	122	29	61
Liberia	158	498	41	287	906	144	453
Mali	15	191	–	39	484	19	242
Niger	10	13	8	20	25	10	13
Nigeria	125	1 386	11	253	2 803	126	1 402
Senegal	37	324	63	85	741	43	371
Sierra Leone	–	–	–	–	–	–	–
Togo	–	–	–	–	–	–	–
Total West Africa	**91**	**6 254**		**175**	**12 038**	**85**	**5 876**
Total Africa	**102**	**63 858**		**191**	**120 137**	**95**	**59 927**
China	67	13 255	92	62	12 191	31	6 096
Democratic People's Republic of Korea	64	395	–	75	464	38	232
Japan	171	4 249	–	152	3 785	76	1 892
Mongolia	131	1 342	46	112	1 148	56	574
Republic of Korea	80	502	54	82	515	41	258
Total East Asia	**81**	**19 743**		**74**	**18 103**	**37**	**9 052**
American Samoa	104	2	–	219	4	110	2
Australia	–	–	–	113	18 510	51	8 339
Cook Islands	–	–	–	–	–	–	–
Fiji	–	–	–	–	–	–	–

TABLE 3 (CONT.)
Forest growing stock, biomass and carbon

Country/area	Growing stock			Biomass		Carbon in biomass	
	Per hectare	Total	Commercial	Per hectare	Total	Per hectare	Total
	(m³/ha)	(million m³)	(%)	(tonnes/ha)	(million tonnes)	(tonnes/ha)	(million tonnes)
French Polynesia	–	–	–	–	–	–	–
Guam	–	–	–	–	–	–	–
Kiribati	–	–	–	–	–	–	–
Marshall Islands	–	–	–	–	–	–	–
Micronesia (Federated States of)	–	–	–	–	–	–	–
Nauru	–	–	–	–	–	–	–
New Caledonia	55	40	58	204	147	102	73
New Zealand	–	–	–	–	–	–	–
Niue	–	–	–	–	–	–	–
Northern Mariana Islands	–	–	–	–	–	–	–
Palau	–	–	–	–	–	–	–
Papua New Guinea	35	1 035	51	–	–	–	–
Pitcairn Islands	–	–	–	–	–	–	–
Samoa	–	–	–	–	–	–	–
Solomon Islands	–	–	–	–	–	–	–
Tokelau	–	–	–	–	–	–	–
Tonga	–	–	–	–	–	–	–
Tuvalu	–	–	–	–	–	–	–
Vanuatu	–	–	–	–	–	–	–
Wallis and Futuna Islands	–	–	–	–	–	–	–
Total Oceania	**36**	**1 077**		**113**	**18 661**	**51**	**8 414**
Bangladesh	34	30	75	72	63	36	31
Bhutan	194	621	40	216	690	108	345
India	69	4 698	40	76	5 178	35	2 343
Maldives	–	–	–	–	–	–	–
Nepal	178	647	40	267	969	133	485
Pakistan	97	185	43	271	516	136	259
Sri Lanka	22	42	40	41	79	21	40
Total South Asia	**79**	**6 223**		**95**	**7 495**	**44**	**3 503**
Brunei Darussalam	219	61	40	283	79	141	39
Cambodia	96	998	40	242	2 532	121	1 266
Indonesia	59	5 216	–	133	11 793	67	5 897
Lao People's Democratic Republic	59	957	74	184	2 974	92	1 487
Malaysia	251	5 242	–	336	7 020	168	3 510
Myanmar	85	2 740	18	197	6 335	98	3 168
Philippines	174	1 248	4	271	1 942	136	971
Singapore	–	–	–	–	–	–	–

Country/area	Growing stock			Biomass		Carbon in biomass	
	Per hectare	Total	Commercial	Per hectare	Total	Per hectare	Total
	(m³/ha)	(million m³)	(%)	(tonnes/ha)	(million tonnes)	(tonnes/ha)	(million tonnes)
Thailand	41	599	60	99	1 434	49	716
Timor-Leste	–	–	–	–	–	–	–
Viet Nam	66	850	9	182	2 348	91	1 174
Total Southeast Asia	**88**	**17 911**		**180**	**36 457**	**90**	**18 228**
Total Asia and the Pacific	**81**	**44 953**		**117**	**80 716**	**57**	**39 197**
Belarus	179	1 411	83	137	1 079	68	539
Republic of Moldova	141	47	62	80	26	40	13
Russian Federation	100	80 479	49	80	64 419	40	32 210
Ukraine	221	2 119	64	156	1 489	78	745
Total CIS countries	**102**	**84 056**		**81**	**67 014**	**41**	**33 507**
Albania	99	79	81	131	104	65	52
Bosnia and Herzegovina	179	391	80	161	351	80	176
Bulgaria	157	568	61	145	527	73	263
Croatia	165	352	83	180	385	90	192
Czech Republic	278	736	97	274	725	123	326
Estonia	196	447	94	146	334	73	167
Hungary	171	337	98	172	340	88	173
Latvia	204	599	85	157	462	79	231
Lithuania	190	400	86	123	258	61	129
Poland	203	1 864	94	195	1 791	97	896
Romania	212	1 347	98	178	1 133	89	567
Serbia and Montenegro	121	327	–	116	312	58	156
Slovakia	256	494	85	211	407	105	203
Slovenia	283	357	91	233	294	116	147
The former Yugoslav Republic of Macedonia	70	63	–	45	41	22	20
Total Eastern Europe	**194**	**8 361**		**173**	**7 463**	**86**	**3 698**
Andorra	–	–	–	–	–	–	–
Austria	300	1 159	98	–	–	–	–
Belgium	258	172	100	196	131	98	65
Channel Islands	–	–	–	–	–	–	–
Denmark	153	77	76	104	52	52	26
Faeroe Islands	–	–	–	–	–	–	–
Finland	96	2 158	84	73	1 631	36	816
France	158	2 465	93	158	2 452	75	1 165
Germany	–	–	–	235	2 605	118	1 303
Gibraltar	–	–	–	–	–	–	–
Greece	47	177	88	31	117	16	59
Holy See	–	–	–	–	–	–	–

TABLE 3 (CONT.)
Forest growing stock, biomass and carbon

Country/area	Growing stock			Biomass		Carbon in biomass	
	Per hectare	Total	Commercial	Per hectare	Total	Per hectare	Total
	(m³/ha)	(million m³)	(%)	(tonnes/ha)	(million tonnes)	(tonnes/ha)	(million tonnes)
Iceland	65	3	–	67	3	33	2
Ireland	98	65	–	59	40	30	20
Isle of Man	–	–	–	–	–	–	–
Italy	145	1 447	70	127	1 272	64	636
Liechtenstein	254	2	80	148	1	74	1
Luxembourg	299	26	100	230	20	103	9
Malta	231	0	0	346	0	173	0
Monaco	–	–	–	–	–	–	–
Netherlands	178	65	80	142	52	68	25
Norway	92	863	78	74	690	37	344
Portugal	93	350	66	60	228	30	114
San Marino	–	–	–	–	–	–	–
Spain	50	888	78	49	871	22	392
Sweden	115	3 155	77	85	2 340	43	1 170
Switzerland	368	449	82	252	308	126	154
United Kingdom	120	340	88	79	224	39	112
Total Western Europe	**115**	**13 861**		**102**	**13 037**	**50**	**6 411**
Total Europe	**107**	**106 278**		**88**	**87 514**	**44**	**43 616**
Anguilla	–	–	–	–	–	–	–
Antigua and Barbuda	–	–	–	–	–	–	–
Aruba	–	–	–	–	–	–	–
Bahamas	13	7	–	–	–	–	–
Barbados	–	–	–	–	–	–	–
Bermuda	–	–	–	–	–	–	–
British Virgin Islands	–	–	–	–	–	–	–
Cayman Islands	–	–	–	–	–	–	–
Cuba	90	243	79	273	740	128	347
Dominica	–	–	–	–	–	–	–
Dominican Republic	47	64	–	119	164	60	82
Grenada	–	–	–	–	–	–	–
Guadeloupe	–	–	–	–	–	–	–
Haiti	65	7	–	158	17	79	8
Jamaica	156	53	2	201	68	100	34
Martinique	–	–	–	–	–	–	–
Montserrat	–	–	–	–	–	–	–
Netherlands Antilles	–	–	–	–	–	–	–
Puerto Rico	65	26	–	103	42	52	21
Saint Kitts and Nevis	–	–	–	–	–	–	–

Country/area	Growing stock			Biomass		Carbon in biomass	
	Per hectare	Total	Commercial	Per hectare	Total	Per hectare	Total
	(m³/ha)	(million m³)	(%)	(tonnes/ha)	(million tonnes)	(tonnes/ha)	(million tonnes)
Saint Lucia	–	–	–	–	–	–	–
Saint Vincent and the Grenadines	–	–	–	–	–	–	–
Trinidad and Tobago	88	20	55	209	47	104	24
Turks and Caicos Islands	–	–	–	–	–	–	–
United States Virgin Islands	15	0	–	57	1	28	0
Total Caribbean	**74**	**420**		**208**	**1 078**	**100**	**516**
Belize	96	159	–	72	118	36	59
Costa Rica	104	249	66	161	385	81	193
El Salvador	–	–	–	–	–	–	–
Guatemala	163	642	15	253	996	126	498
Honduras	116	540	–	–	–	–	–
Nicaragua	114	591	25	276	1 432	138	716
Panama	160	686	1	288	1 238	144	620
Total Central America	**130**	**2 867**		**239**	**4 169**	**119**	**2 086**
Argentina	55	1 826	67	146	4 817	73	2 411
Bolivia	74	4 360	16	180	10 568	90	5 296
Brazil	170	81 239	18	212	101 236	103	49 335
Chile	117	1 882	64	241	3 892	121	1 946
Colombia	–	–	–	266	16 125	133	8 062
Ecuador	–	–	–	–	–	–	–
Falkland Islands	–	–	–	–	–	–	–
French Guiana	350	2 822	0	–	–	–	–
Guyana	–	–	–	228	3 443	114	1 722
Paraguay	–	–	–	–	–	–	–
Peru	–	–	–	–	–	–	–
South Georgia and the South Sandwich Islands	–	–	–	–	–	–	–
Suriname	150	2 216	–	770	11 383	385	5 692
Uruguay	79	118	6	–	–	–	–
Venezuela (Bolivarian Republic of)	–	–	–	–	–	–	–
Total South America	**155**	**94 464**		**224**	**151 464**	**110**	**74 464**
Total Latin America and the Caribbean	**153**	**97 751**		**224**	**156 711**	**110**	**77 066**
Canada	106	32 983	100	–	–	–	–
Greenland	–	–	–	–	–	–	–
Mexico	–	–	–	–	–	–	–
Saint Pierre and Miquelon	–	–	–	–	–	–	–
United States of America	116	35 118	79	125	37 929	63	18 964
Total North America	**111**	**68 101**		**125**	**37 929**	**63**	**18 964**

TABLE 3 (CONT.)
Forest growing stock, biomass and carbon

Country/area	Growing stock			Biomass		Carbon in biomass	
	Per hectare	Total	Commercial	Per hectare	Total	Per hectare	Total
	(m³/ha)	(million m³)	(%)	(tonnes/ha)	(million tonnes)	(tonnes/ha)	(million tonnes)
Armenia	125	36	–	128	36	64	18
Azerbaijan	136	127	20	124	116	62	58
Georgia	167	461	26	152	420	76	210
Kazakhstan	109	364	0	82	273	41	137
Kyrgyzstan	34	30	0	29	25	14	13
Tajikistan	12	5	0	14	6	7	3
Turkmenistan	4	15	0	8	35	4	17
Uzbekistan	7	24	0	7	25	4	12
Total Central Asia	**66**	**1 061**		**58**	**935**	**29**	**468**
Afghanistan	16	14	40	15	13	7	6
Bahrain	–	–	–	–	–	–	–
Cyprus	46	8	39	32	6	16	3
Iran (Islamic Republic of)	48	527	79	60	669	30	334
Iraq	–	–	–	–	–	–	–
Israel	37	6	70	–	–	–	–
Jordan	30	3	–	56	5	28	2
Kuwait	–	–	–	–	–	–	–
Lebanon	36	5	–	26	4	13	2
Occupied Palestinian Territory	–	–	–	–	–	–	–
Oman	–	–	–	–	–	–	–
Qatar	–	–	–	–	–	–	–
Saudi Arabia	8	23	0	13	35	6	17
Syrian Arab Republic	–	–	–	–	–	–	–
Turkey	138	1 400	87	161	1 634	80	817
United Arab Emirates	49	15	0	106	33	53	17
Yemen	9	5	–	19	10	9	5
Total Western Asia	**76**	**2 006**		**92**	**2 407**	**46**	**1 203**
Total Western and Central Asia	**73**	**3 067**		**79**	**3 343**	**40**	**1 671**
TOTAL WORLD	**111**	**384 007**		**145**	**486 350**	**72**	**240 441**

SOURCE: FAO, 2006a.

TABLE 4
Production, trade and consumption of woodfuel, roundwood and sawnwood, 2006

Country/area	Woodfuel				Industrial roundwood				Sawnwood			
	(1 000 m³)				(1 000 m³)				(1 000 m³)			
	Production	Imports	Exports	Consumption	Production	Imports	Exports	Consumption	Production	Imports	Exports	Consumption
Burundi	8 681	0	0	8 681	333	0	7	326	83	0	0	83
Cameroon	9 566	0	0	9 566	1 800	0	29	1 771	702	0	514	188
Central African Republic	2 000	0	0	2 000	832	0	85	747	69	0	11	58
Chad	6 600	0	0	6 600	761	0	0	761	2	18	1	19
Congo	1 256	0	0	1 256	2 331	0	633	1 698	268	0	181	87
Democratic Republic of the Congo	72 126	0	0	72 126	4 322	1	89	4 234	94	1	69	26
Equatorial Guinea	447	0	0	447	700	0	685	15	7	0	6	1
Gabon	530	0	0	530	3 500	0	1 787	1 713	235	0	199	36
Rwanda	9 416	0	0	9 416	495	0	0	495	79	0	0	79
Saint Helena	0	0	0	0	0	0	0	0	0	0	0	0
Sao Tome and Principe	0	0	0	0	9	0	0	9	5	0	1	5
Total Central Africa	**110 621**	**0**	**0**	**110 621**	**15 083**	**2**	**3 316**	**11 768**	**1 544**	**19**	**982**	**582**
British Indian Ocean Territory	0	0	0	0	0	0	0	0	0	0	0	0
Comoros	0	0	0	0	9	0	0	9	0	1	0	1
Djibouti	0	0	0	0	0	1	0	1	0	2	0	2
Eritrea	2 486	0	0	2 486	2	6	0	8	0	0	0	0
Ethiopia	95 703	0	0	95 703	2 928	0	0	2 928	18	10	0	28
Kenya	20 749	0	0	20 749	1 813	8	1	1 820	142	2	1	144
Madagascar	11 339	0	0	11 339	183	0	43	140	89	1	28	62
Mauritius	7	0	0	7	9	20	1	28	4	65	1	68
Mayotte	–	–	–	–	–	–	–	–	–	–	–	–
Réunion	31	0	0	31	5	1	2	3	2	85	0	87
Seychelles	0	0	0	0	0	0	0	0	0	0	0	0
Somalia	11 127	0	0	11 127	110	1	5	106	14	1	0	15
Uganda	37 343	0	0	37 343	3 175	0	0	3 175	117	0	1	116
United Republic of Tanzania	21 914	0	1	21 913	2 314	2	57	2 259	40	1	32	10
Total East Africa	**200 699**	**0**	**1**	**200 698**	**10 547**	**39**	**110**	**10 476**	**427**	**168**	**63**	**533**
Algeria	7 767	0	0	7 767	75	34	1	108	13	1 157	0	1 169
Egypt	17 059	0	0	17 059	268	116	0	384	2	1 463	0	1 465
Libyan Arab Jamahiriya	901	0	0	901	116	8	0	124	31	123	0	154
Mauritania	1 663	0	0	1 663	3	1	0	4	14	0	0	14
Morocco	345	0	0	345	599	462	0	1 061	83	1 043	0	1 126
Sudan	17 901	0	0	17 901	2 173	0	0	2 173	51	58	0	109

TABLE 4 (CONT.)

Production, trade and consumption of woodfuel, roundwood and sawnwood, 2006

Country/area	Woodfuel				Industrial roundwood				Sawnwood			
	(1 000 m³)				(1 000 m³)				(1 000 m³)			
	Production	Imports	Exports	Consumption	Production	Imports	Exports	Consumption	Production	Imports	Exports	Consumption
Tunisia	2 156	0	0	2 156	218	81	0	299	20	562	2	581
Western Sahara	–	–	–	–	–	–	–	–	–	–	–	–
Total Northern Africa	**47 792**	**0**	**0**	**47 792**	**3 452**	**702**	**2**	**4 153**	**214**	**4 407**	**3**	**4 618**
Angola	3 656	0	0	3 656	1 096	2	4	1 093	5	1	0	6
Botswana	665	0	0	665	105	0	0	105	15	0	0	15
Lesotho	2 061	0	0	2 061	0	0	0	0	0	0	0	0
Malawi	5 189	0	0	5 189	520	2	0	521	45	0	16	29
Mozambique	16 724	0	0	16 724	1 304	4	133	1 175	43	19	19	43
Namibia	–	–	–	–	–	–	–	–	–	–	–	–
South Africa	12 000	0	0	12 000	18 063	51	191	17 922	2 091	487	63	2 516
Swaziland	996	0	0	996	330	0	0	330	102	0	0	102
Zambia	8 798	0	0	8 798	1 325	0	1	1 325	157	1	6	153
Zimbabwe	8 380	0	0	8 380	771	1	5	767	565	2	83	484
Total Southern Africa	**58 469**	**0**	**0**	**58 469**	**23 514**	**60**	**334**	**23 239**	**3 023**	**511**	**186**	**3 348**
Benin	6 101	0	0	6 101	332	0	13	319	31	9	18	21
Burkina Faso	12 067	0	0	12 067	1 171	3	3	1 171	1	21	4	17
Cape Verde	2	0	0	2	0	2	0	1	0	1	0	1
Côte d'Ivoire	8 740	0	0	8 740	1 347	10	142	1 215	420	0	381	39
Gambia	656	0	0	656	113	0	0	112	1	2	0	3
Ghana	33 040	0	0	33 040	1 304	3	1	1 305	527	0	210	317
Guinea	11 738	0	0	11 738	651	1	23	629	10	0	9	2
Guinea-Bissau	422	0	0	422	170	0	7	163	16	1	0	16
Liberia	6 033	0	0	6 033	300	0	0	300	60	0	1	59
Mali	5 084	0	0	5 084	413	1	1	413	13	0	0	13
Niger	9 010	0	0	9 010	411	1	4	408	4	0	0	4
Nigeria	61 629	0	1	61 628	9 418	1	42	9 377	2 000	1	22	1 980
Senegal	5 306	0	0	5 306	794	23	0	817	23	86	1	108
Sierra Leone	5 448	0	0	5 448	124	0	1	123	5	1	0	6
Togo	5 816	0	0	5 816	166	0	8	158	14	4	5	14
Total West Africa	**171 091**	**0**	**1**	**171 091**	**16 713**	**44**	**247**	**16 511**	**3 124**	**127**	**651**	**2 599**
Total Africa	**588 673**	**1**	**3**	**588 670**	**69 309**	**847**	**4 009**	**66 147**	**8 332**	**5 233**	**1 885**	**11 679**
China	203 505	18	9	203 514	94 665	33 239	720	127 184	10 245	8 108	846	17 508
Democratic People's Republic of Korea	5 835	0	0	5 835	1 500	0	40	1 460	280	1	22	259
Japan	105	1	0	106	16 609	10 582	32	27 159	12 554	8 505	17	21 042
Mongolia	704	0	0	704	40	7	1	46	30	2	3	29
Republic of Korea	2 469	0	0	2 469	2 444	6 366	0	8 810	4 366	804	15	5 155
Total East Asia	**212 618**	**19**	**9**	**212 628**	**115 258**	**50 194**	**792**	**164 659**	**27 475**	**17 420**	**903**	**43 992**

Country/area	Woodfuel (1 000 m³)				Industrial roundwood (1 000 m³)				Sawnwood (1 000 m³)			
	Production	Imports	Exports	Consumption	Production	Imports	Exports	Consumption	Production	Imports	Exports	Consumption
American Samoa	–	–	–	–	0	0	0	0	0	1	0	1
Australia	6 969	0	0	6 969	26 904	2	1 065	25 841	4 784	570	344	5 010
Cook Islands	0	0	0	0	5	0	1	4	0	3	0	3
Fiji	37	0	0	37	472	2	6	468	125	2	20	107
French Polynesia	–	–	–	–	0	4	0	4	0	40	0	40
Guam	–	–	–	–	–	–	–	–	–	–	–	–
Kiribati	0	0	0	0	0	0	0	0	0	2	0	2
Marshall Islands	–	–	–	–	–	–	–	–	0	6	0	6
Micronesia (Federated States of)	0	0	0	0	0	0	0	0	0	7	0	7
Nauru	0	0	0	0	0	0	0	0	0	0	0	0
New Caledonia	0	0	0	0	5	4	1	8	3	20	1	22
New Zealand	–	0	0	–	19 254	3	5 571	13 687	4 269	50	1 960	2 359
Niue	–	–	–	–	0	0	0	0	0	0	0	0
Northern Mariana Islands	–	–	–	–	–	–	–	–	0	0	0	0
Palau	–	–	–	–	0	1	0	1	0	3	0	3
Papua New Guinea	5 533	0	0	5 533	2 908	0	2 638	270	60	0	51	9
Pitcairn Islands	–	–	–	–	–	–	–	–	0	0	0	0
Samoa	70	0	0	70	61	6	1	66	21	22	0	43
Solomon Islands	138	0	0	138	1 130	0	1 011	119	12	0	11	1
Tokelau	–	–	–	–	–	–	–	–	0	0	0	0
Tonga	0	2	0	2	2	1	2	1	2	14	0	16
Tuvalu	–	–	–	–	0	0	0	0	0	1	0	1
Vanuatu	91	0	1	90	28	2	0	30	28	2	2	28
Wallis and Futuna Islands	–	–	–	–	0	0	0	1				0
Total Oceania	**12 838**	**2**	**1**	**12 839**	**50 769**	**25**	**10 294**	**40 500**	**9 304**	**745**	**2 390**	**7 660**
Bangladesh	27 584	0	0	27 584	282	329	1	611	388	2	0	390
Bhutan	4 546	0	0	4 546	133	0	3	130	31	0	0	31
India	306 252	79	0	306 332	23 192	4 043	3	27 231	14 789	173	19	14 943
Maldives	0	0	0	0	0	0	0	0	0	0	0	0
Nepal	12 654	0	0	12 654	1 260	1	2	1 259	630	2	0	631
Pakistan	26 124	0	0	26 124	2 870	259	0	3 129	1 313	120	0	1 433
Sri Lanka	5 584	0	0	5 584	694	1	3	693	61	30	0	90
Total South Asia	**382 745**	**80**	**0**	**382 825**	**28 431**	**4 634**	**12**	**33 053**	**17 212**	**326**	**19**	**17 519**
Brunei Darussalam	12	0	0	12	112	0	0	112	51	0	1	50
Cambodia	9 221	0	0	9 221	113	1	0	114	74	0	63	11
Indonesia	70 719	0	1	70 718	28 099	120	685	27 534	3 853	311	1 853	2 311
Lao People's Democratic Republic	5 944	0	0	5 944	194	0	63	131	140	0	131	9
Malaysia	3 013	12	0	3 024	22 506	138	4 909	17 735	5 129	1 004	2 608	3 525

TABLE 4 (CONT.)
Production, trade and consumption of woodfuel, roundwood and sawnwood, 2006

Country/area	Woodfuel				Industrial roundwood				Sawnwood			
	(1 000 m³)				(1 000 m³)				(1 000 m³)			
	Production	Imports	Exports	Consumption	Production	Imports	Exports	Consumption	Production	Imports	Exports	Consumption
Myanmar	38 286	0	0	38 286	4 262	0	1 476	2 786	1 530	0	275	1 256
Philippines	12 821	0	0	12 821	2 927	138	7	3 058	468	264	184	548
Singapore	0	1	0	1	0	40	39	1	25	224	195	54
Thailand	19 736	0	0	19 736	8 700	398	0	9 098	288	1 890	1 314	864
Timor-Leste	–	–	–	–	0	0	0	0	0	0	0	0
Viet Nam	26 151	0	0	26 151	4 678	203	8	4 873	3 414	531	81	3 864
Total Southeast Asia	**185 903**	**13**	**1**	**185 915**	**71 590**	**1 039**	**7 188**	**65 442**	**14 972**	**4 225**	**6 704**	**12 493**
Total Asia and the Pacific	**794 104**	**114**	**11**	**794 207**	**266 048**	**55 891**	**18 286**	**303 654**	**68 964**	**22 716**	**10 016**	**81 664**
Belarus	1 345	1	75	1 271	7 411	76	1 443	6 044	2 458	116	1 197	1 377
Republic of Moldova	94	2	0	96	94	28	0	122	31	110	0	141
Russian Federation	46 000	0	200	45 800	144 600	516	50 900	94 216	22 127	15	15 900	6 242
Ukraine	8 494	1	498	7 997	6 752	173	2 202	4 723	2 192	9	1 249	952
Total CIS countries	**55 933**	**3**	**772**	**55 164**	**158 857**	**793**	**54 545**	**105 105**	**26 808**	**249**	**18 345**	**8 712**
Albania	221	0	56	165	75	1	0	75	97	24	21	99
Bosnia and Herzegovina	1 459	0	290	1 169	2 646	53	156	2 544	1 319	17	932	404
Bulgaria	2 885	0	147	2 738	3 107	46	581	2 572	569	28	269	329
Croatia	915	3	295	623	3 537	67	612	2 992	669	371	477	563
Czech Republic	1 345	48	280	1 113	16 333	1 225	2 679	14 879	5 080	507	2 000	3 587
Estonia	1 100	2	51	1 051	4 300	1 809	1 606	4 503	1 923	753	970	1 705
Hungary	3 246	168	214	3 200	2 667	189	1 095	1 761	186	852	172	866
Latvia	979	2	405	576	11 866	1 216	3 419	9 663	4 320	481	2 572	2 229
Lithuania	1 230	13	83	1 160	4 640	197	1 061	3 777	1 466	538	803	1 200
Montenegro	265	0	30	235	192	1	44	149	77	2	49	30
Poland	3 617	14	78	3 553	28 767	1 814	412	30 169	3 607	541	603	3 545
Romania	4 516	1	79	4 438	9 454	425	111	9 768	3 476	48	2 351	1 173
Serbia	1 626	1	2	1 625	1 250	87	48	1 289	493	419	144	768
Slovakia	307	10	15	302	7 562	340	1 218	6 684	2 440	72	1 192	1 320
Slovenia	984	58	175	867	2 195	363	383	2 175	580	223	433	370
The former Yugoslav Republic of Macedonia	662	0	5	657	162	1	6	158	17	52	9	60
Total Eastern Europe	**25 357**	**320**	**2 204**	**23 473**	**98 753**	**7 834**	**13 430**	**93 157**	**26 319**	**4 927**	**12 997**	**18 249**
Andorra	0	2	0	2	0	0	0	0	0	10	0	10
Austria	4 705	326	54	4 977	14 430	9 102	718	22 814	10 507	1 881	6 889	5 499
Belgium	670	45	9	707	4 405	3 284	1 025	6 664	1 520	2 213	1 065	2 668
Channel Islands	–	–	–	–	–	–	–	–	–	–	–	–
Denmark	1 162	305	37	1 430	1 196	848	645	1 399	196	2 201	143	2 253

Country/area	Woodfuel				Industrial roundwood				Sawnwood			
	(1 000 m³)				(1 000 m³)				(1 000 m³)			
	Production	Imports	Exports	Consumption	Production	Imports	Exports	Consumption	Production	Imports	Exports	Consumption
Faeroe Islands	0	0	0	0	0	1	0	1	0	4	0	4
Finland	5 290	174	9	5 455	45 521	14 655	709	59 468	12 227	578	7 728	5 077
France	33 198	44	560	32 682	28 592	2 601	3 695	27 498	9 992	3 922	1 493	12 421
Germany	8 290	547	79	8 759	54 000	3 669	7 557	50 113	24 420	5 307	8 789	20 938
Gibraltar	0	0	0	0	0	0	0	0	0	1	0	1
Greece	1 004	69	7	1 066	519	190	4	705	191	948	12	1 127
Holy See	–	–	–	–	–	–	–	–	–	–	–	–
Iceland	0	0	0	0	0	1	0	1	0	95	0	95
Ireland	16	1	1	16	2 655	208	308	2 555	1 094	995	393	1 697
Isle of Man	–	–	–	–	–	–	–	–	–	–	–	–
Italy	5 606	1 099	2	6 703	3 013	4 486	15	7 484	1 748	7 862	169	9 441
Liechtenstein	4	0	0	4	18	0	0	18	–	–	–	–
Luxembourg	–	20	35	–	255	351	224	383	133	57	38	152
Malta	0	0	0	0	0	0	0	0	0	19	0	19
Monaco	–	–	–	–	–	–	–	–	–	–	–	–
Netherlands	290	2	30	261	817	390	570	636	265	3 399	555	3 109
Norway	1 177	175	5	1 347	7 417	2 334	740	9 011	2 389	1 035	474	2 950
Portugal	600	2	8	594	10 205	335	1 422	9 118	1 010	258	462	806
San Marino	–	–	–	–	–	–	–	–	–	–	–	–
Spain	1 607	42	188	1 461	14 109	3 841	224	17 726	3 806	3 373	117	7 062
Sweden	5 900	230	42	6 088	58 700	6 664	3 004	62 360	18 300	384	13 217	5 467
Switzerland	1 417	8	37	1 388	4 285	346	1 727	2 904	1 668	409	252	1 825
United Kingdom	317	4	145	176	8 100	415	644	7 871	2 902	7 963	415	10 449
Total Western Europe	**71 255**	**3 095**	**1 247**	**73 118**	**258 235**	**53 722**	**23 229**	**288 729**	**92 369**	**42 912**	**42 211**	**93 070**
Total Europe	**152 544**	**3 418**	**4 223**	**151 755**	**515 845**	**62 349**	**91 204**	**486 991**	**145 496**	**48 088**	**73 554**	**120 030**
Anguilla	–	–	–	–	–	–	–	–	–	–	–	–
Antigua and Barbuda	–	–	–	–	0	0	0	0	0	11	0	11
Aruba	0	0	0	0	0	1	0	1	0	16	0	16
Bahamas	0	1	0	1	17	63	0	80	1	108	2	107
Barbados	0	3	0	3	6	5	0	11	0	24	0	24
Bermuda	–	–	–	–	–	–	–	–	–	–	–	–
British Virgin Islands	0	0	0	0	0	0	0	0	0	4	0	4
Cayman Islands	0	0	0	0	0	2	0	2	0	14	0	14
Cuba	1 584	0	0	1 584	761	0	0	761	243	8	0	251
Dominica	0	0	0	0	0	1	0	1	66	4	2	67
Dominican Republic	878	0	0	878	14	17	0	30	12	310	0	322
Grenada	0	0	0	0	0	0	0	0	0	10	0	10
Guadeloupe	32	0	0	32	0	5	0	5	1	46	0	47
Haiti	2 008	0	0	2 008	239	1	0	240	14	19	0	33
Jamaica	559	0	0	560	278	3	0	281	66	38	0	104
Martinique	25	0	0	25	2	3	0	5	1	29	0	30
Montserrat	0	0	0	0	0	0	0	0	0	4	0	4

TABLE 4 (CONT.)

Production, trade and consumption of woodfuel, roundwood and sawnwood, 2006

Country/area	Woodfuel				Industrial roundwood				Sawnwood			
	(1 000 m³)				(1 000 m³)				(1 000 m³)			
	Production	Imports	Exports	Consumption	Production	Imports	Exports	Consumption	Production	Imports	Exports	Consumption
Netherlands Antilles	0	0	0	0	0	1	0	1	0	20	0	20
Puerto Rico	–	–	–	–	–	–	–	–	–	–	–	–
Saint Kitts and Nevis	0	0	0	0	0	1	0	1	0	5	0	5
Saint Lucia	0	0	0	0	0	7	0	7	0	15	0	15
Saint Vincent and the Grenadines	0	0	0	0	0	2	0	2	0	12	0	12
Trinidad and Tobago	34	0	0	34	65	5	1	70	41	40	0	81
Turks and Caicos Islands	0	0	0	0	0	0	0	0	0	4	0	4
United States Virgin Islands	–	–	–	–	–	–	–	–	–	–	–	–
Total Caribbean	**5 120**	**5**	**0**	**5 125**	**1 382**	**117**	**1**	**1 498**	**445**	**738**	**5**	**1 178**
Belize	126	0	0	126	62	2	0	63	35	9	2	42
Costa Rica	3 424	0	0	3 423	1 198	3	62	1 139	488	29	3	514
El Salvador	4 204	0	0	4 204	682	2	2	683	16	53	0	69
Guatemala	16 609	0	0	16 609	454	1	18	437	366	3	53	316
Honduras	8 668	0	1	8 667	873	5	68	811	400	17	91	326
Nicaragua	5 975	1	0	5 975	93	1	7	87	54	0	50	4
Panama	1 189	0	0	1 189	160	6	80	86	30	10	19	21
Total Central America	**40 195**	**1**	**1**	**40 194**	**3 522**	**21**	**237**	**3 305**	**1 390**	**121**	**218**	**1 293**
Argentina	4 372	0	0	4 372	9 846	2	35	9 813	1 739	114	384	1 468
Bolivia	2 270	0	0	2 270	810	1	2	809	408	4	59	353
Brazil	138 783	0	0	138 783	100 767	34	121	100 680	23 557	134	3 167	20 524
Chile	13 899	0	0	13 899	33 217	0	111	33 106	8 718	31	3 391	5 358
Colombia	10 350	0	0	10 350	1 637	0	10	1 627	389	9	5	393
Ecuador	5 574	0	0	5 574	1 211	0	47	1 165	755	0	37	719
Falkland Islands	0	0	0	0	0	0	0	0	0	0	0	0
French Guiana	105	0	0	105	66	1	2	65	15	1	4	12
Guyana	860	0	0	860	574	0	150	424	68	0	36	32
Paraguay	6 149	0	0	6 149	4 044	0	13	4 031	550	41	44	547
Peru	7 454	0	0	7 454	1 804	3	0	1 807	856	26	172	710
South Georgia and the South Sandwich Islands	–	–	–	–	–	–	–	–	–	–	–	–
Suriname	45	0	0	45	194	0	1	193	69	0	0	69
Uruguay	2 111	0	0	2 111	3 885	8	1 996	1 897	268	30	130	168
Venezuela (Bolivarian Republic of)	3 884	0	0	3 884	1 673	0	6	1 667	838	31	6	863
Total South America	**195 856**	**0**	**0**	**195 856**	**159 728**	**50**	**2 493**	**157 284**	**38 230**	**421**	**7 435**	**31 216**
Total Latin America and the Caribbean	**241 171**	**5**	**1**	**241 175**	**164 631**	**187**	**2 732**	**162 087**	**40 065**	**1 280**	**7 658**	**33 687**

Country/area	Woodfuel				Industrial roundwood				Sawnwood			
	(1 000 m³)				*(1 000 m³)*				*(1 000 m³)*			
	Production	Imports	Exports	Consumption	Production	Imports	Exports	Consumption	Production	Imports	Exports	Consumption
Canada	2 997	90	218	2 869	185 196	5 787	4 640	186 343	58 709	1 546	38 984	21 271
Greenland	0	0	0	0	0	1	0	1	0	7	0	7
Mexico	38 521	2	7	38 516	6 193	174	9	6 358	2 829	4 193	64	6 958
Saint Pierre and Miquelon	0	0	0	0	0	0	0	0	0	2	0	2
United States of America	44 914	170	135	44 949	412 134	2 922	9 638	405 418	92 903	40 109	4 607	128 406
Total North America	**86 432**	**262**	**360**	**86 334**	**603 523**	**8 883**	**14 287**	**598 120**	**154 442**	**45 857**	**43 655**	**156 644**
Armenia	60	0	0	60	5	3	2	6	5	50	2	53
Azerbaijan	3	0	0	3	3	21	0	24	0	578	1	577
Georgia	454	0	0	454	162	0	3	159	150	0	130	20
Kazakhstan	210	5	0	215	642	171	0	813	139	813	127	825
Kyrgyzstan	18	0	0	18	9	4	0	13	22	107	2	127
Tajikistan	90	0	0	90	0	0	0	0	0	109	0	109
Turkmenistan	3	0	0	3	0	0	0	0	0	24	0	24
Uzbekistan	22	0	0	22	9	373	4	377	0	1	1	1
Total Central Asia	**861**	**5**	**0**	**865**	**831**	**571**	**9**	**1 392**	**316**	**1 681**	**262**	**1 735**
Afghanistan	1 498	0	0	1 498	1 760	2	10	1 752	400	258	0	658
Bahrain	0	0	0	0	0	1	0	1	0	138	0	138
Cyprus	3	0	0	3	5	0	0	5	4	120	0	124
Iran (Islamic Republic of)	65	1	0	66	729	97	0	826	50	760	0	810
Iraq	57	0	0	57	59	1	0	60	12	69	0	81
Israel	2	0	0	2	25	140	0	164	0	454	0	454
Jordan	269	0	0	269	4	7	2	10	0	256	7	249
Kuwait	0	0	0	0	0	7	0	7	0	129	0	129
Lebanon	81	0	0	81	7	38	1	45	9	248	1	256
Occupied Palestinian Territory	–	–	–	–	–	–	–	–	–	–	–	–
Oman	0	0	0	0	0	57	0	57	0	83	0	82
Qatar	0	10	0	10	0	34	3	31	0	80	0	80
Saudi Arabia	0	4	0	4	0	25	0	25	0	1 599	0	1 599
Syrian Arab Republic	25	0	18	7	40	4	0	43	9	572	0	581
Turkey	5 831	233	0	6 064	12 253	2 022	3	14 272	6 471	626	44	7 053
United Arab Emirates	0	0	0	0	0	160	3	156	0	484	12	472
Yemen	381	0	0	381	0	10	0	10	0	160	0	160
Total Western Asia	**8 212**	**249**	**18**	**8 443**	**14 882**	**2 607**	**23**	**17 466**	**6 955**	**6 035**	**65**	**12 925**
Total Western and Central Asia	**9 072**	**254**	**18**	**9 308**	**15 713**	**3 178**	**32**	**18 859**	**7 271**	**7 716**	**327**	**14 660**
TOTAL WORLD	**1 871 996**	**4 055**	**4 617**	**1 871 450**	**1 635 069**	**131 336**	**130 549**	**1 635 857**	**424 568**	**130 890**	**137 094**	**418 364**

SOURCE: FAOSTAT (ForesSTAT), last accessed 28 August 2008.

TABLE 5

Production, trade and consumption of wood-based panels, pulp and paper, 2006

Country/area	Wood-based panels				Pulp for paper				Paper and paperboard			
	(1 000 m³)				(1 000 m³)				(1 000 m³)			
	Production	Imports	Exports	Consumption	Production	Imports	Exports	Consumption	Production	Imports	Exports	Consumption
Burundi	0	1	0	0	0	0	0	0	0	1	0	1
Cameroon	88	0	51	37	0	0	0	0	0	39	0	39
Central African Republic	2	0	0	2	0	0	0	0	0	1	1	0
Chad	0	1	0	1	0	0	0	0	0	0	0	0
Congo	20	0	6	14	0	0	0	0	0	5	0	5
Democratic Republic of the Congo	3	1	1	2	0	0	0	0	0	10	1	10
Equatorial Guinea	30	1	26	5	0	0	0	0	0	0	0	0
Gabon	292	0	277	15	0	0	0	0	0	5	0	5
Rwanda	0	1	0	1	0	0	0	0	0	4	0	3
Saint Helena	0	0	0	0	0	0	0	0	–	–	–	–
Sao Tome and Principe	0	0	0	0	0	0	0	0	–	–	–	–
Total Central Africa	**434**	**5**	**361**	**78**	**0**	**2**	**0**	**1**	**0**	**65**	**2**	**63**
British Indian Ocean Territory	0	0	0	0	0	0	0	0	0	0	0	0
Comoros	0	0	0	0	0	0	0	0	–	–	–	–
Djibouti	0	11	0	11	0	3	0	3	0	9	0	8
Eritrea	0	0	0	0	0	0	0	0	0	2	0	2
Ethiopia	83	2	0	85	9	2	0	12	16	17	0	33
Kenya	83	13	5	91	113	2	0	115	234	124	15	343
Madagascar	5	5	0	9	0	3	0	3	10	20	0	29
Mauritius	0	61	3	57	0	2	0	2	0	48	3	44
Mayotte	–	–	–	–	–	–	–	–	–	–	–	–
Réunion	0	24	0	23	0	0	0	0	0	15	0	15
Seychelles	0	1	0	1	0	0	0	0	–	–	–	–
Somalia	0	0	0	0	0	0	0	0	–	–	–	–
Uganda	24	8	4	28	0	0	0	0	3	44	1	46
United Republic of Tanzania	5	24	1	28	56	0	0	56	25	102	4	123
Total East Africa	**199**	**148**	**14**	**333**	**178**	**13**	**0**	**192**	**288**	**380**	**24**	**644**
Algeria	48	49	0	97	2	4	0	6	35	236	0	270
Egypt	56	364	1	419	120	105	0	225	460	748	47	1 161
Libyan Arab Jamahiriya	0	26	0	26	0	4	0	4	0	35	0	35
Mauritania	2	0	0	2	0	0	0	0	0	3	0	3
Morocco	35	117	27	126	112	23	123	12	129	255	11	373
Sudan	2	47	0	49	0	0	0	0	3	39	0	41
Tunisia	104	84	22	165	10	97	12	95	106	215	52	268
Western Sahara	–	–	–	–	–	–	–	–	–	–	–	–
Total Northern Africa	**247**	**688**	**50**	**885**	**244**	**233**	**135**	**342**	**732**	**1 530**	**111**	**2 151**

Country/area	Wood-based panels (1 000 m³)				Pulp for paper (1 000 m³)				Paper and paperboard (1 000 m³)			
	Production	Imports	Exports	Consumption	Production	Imports	Exports	Consumption	Production	Imports	Exports	Consumption
Angola	11	4	0	15	15	0	0	15	0	12	0	11
Botswana	0	0	0	0	0	0	0	0	0	10	0	10
Lesotho	0	0	0	0	0	0	0	0	–	–	–	–
Malawi	18	3	6	15	0	0	0	0	0	19	0	19
Mozambique	3	5	2	7	0	0	1	0	0	12	0	12
Namibia	–	–	–	–	–	–	–	–	–	–	–	–
South Africa	726	355	75	1 007	2 915	515	972	2 457	1 793	59	210	1 642
Swaziland	8	0	0	8	167	0	167	0	–	–	–	–
Zambia	18	4	4	18	0	0	0	0	4	27	0	31
Zimbabwe	80	15	19	76	49	10	0	59	115	45	13	146
Total Southern Africa	**864**	**386**	**105**	**1 146**	**3 146**	**525**	**1 140**	**2 531**	**1 912**	**183**	**224**	**1 871**
Benin	0	2	0	2	0	0	0	0	0	6	0	6
Burkina Faso	0	2	0	2	0	0	0	0	0	11	0	11
Cape Verde	0	1	0	0	0	0	0	0	0	2	0	2
Côte d'Ivoire	301	0	232	69	0	0	0	0	0	71	2	69
Gambia	0	2	1	1	0	0	0	0	–	–	–	–
Ghana	335	1	175	161	0	0	0	0	0	65	0	65
Guinea	42	2	3	41	0	0	0	0	0	3	0	3
Guinea-Bissau	0	0	0	0	0	0	0	0	0	0	0	0
Liberia	0	5	0	4	0	0	0	0	0	2	0	2
Mali	0	0	0	0	0	0	0	0	0	5	0	5
Niger	0	0	0	0	0	8	0	8	0	1	0	1
Nigeria	95	42	0	136	23	17	0	40	19	297	2	315
Senegal	0	11	0	11	0	0	0	0	0	31	2	29
Sierra Leone	0	3	1	3	0	0	0	0	0	1	1	0
Togo	0	1	0	1	0	0	0	0	0	5	0	5
Total West Africa	**773**	**73**	**413**	**433**	**23**	**26**	**0**	**49**	**19**	**500**	**8**	**511**
Total Africa	**2 517**	**1 300**	**943**	**2 874**	**3 591**	**801**	**1 276**	**3 116**	**2 951**	**2 658**	**369**	**5 240**
China	63 842	4 941	9 774	59 010	18 976	8 178	114	27 040	57 983	8 636	5 683	60 936
Democratic People's Republic of Korea	0	9	0	9	106	45	0	151	80	25	2	102
Japan	5 514	5 646	33	11 127	10 847	2 211	210	12 848	29 473	1 959	1 456	29 976
Mongolia	2	4	1	5	0	0	0	0	0	5	0	5
Republic of Korea	3 760	2 962	47	6 675	516	2 422	0	2 938	11 040	768	3 165	8 643
Total East Asia	**73 118**	**13 562**	**9 855**	**76 826**	**30 445**	**12 856**	**324**	**42 977**	**98 576**	**11 392**	**10 306**	**99 663**
American Samoa	0	0	0	0	0	0	0	0	0	0	0	0
Australia	1 989	394	422	1 961	1 153	344	10	1 487	3 221	1 551	808	3 964
Cook Islands	0	2	0	2	0	0	0	0	0	0	0	0
Fiji	16	16	3	29	0	0	0	0	0	26	1	25
French Polynesia	0	10	0	9	0	0	0	0	0	7	0	7

TABLE 5 (CONT.)

Production, trade and consumption of wood-based panels, pulp and paper, 2006

Country/area	Wood-based panels (1 000 m³)				Pulp for paper (1 000 m³)				Paper and paperboard (1 000 m³)			
	Production	Imports	Exports	Consumption	Production	Imports	Exports	Consumption	Production	Imports	Exports	Consumption
Guam	–	–	–	–	–	–	–	–	–	–	–	–
Kiribati	0	0	0	0	0	0	0	0	0	0	0	0
Marshall Islands	0	3	0	3	0	0	0	0	0	0	0	0
Micronesia (Federated States of)	0	1	0	1	0	0	0	0	0	0	0	0
Nauru	0	0	0	0	0	0	0	0	0	0	0	0
New Caledonia	0	9	3	6	0	0	0	0	0	9	7	1
New Zealand	2 223	46	1 043	1 226	1 562	5	699	868	944	470	593	821
Niue	0	0	0	0	0	0	0	0	0	0	0	0
Northern Mariana Islands	0	0	0	0	0	0	0	0	0	0	0	0
Palau	0	1	0	1	0	0	0	0	0	0	0	0
Papua New Guinea	88	1	68	21	0	0	0	0	0	16	0	16
Pitcairn Islands	0	0	0	0	0	0	0	0	0	0	0	0
Samoa	0	2	0	2	0	0	0	0	0	1	0	1
Solomon Islands	0	0	0	0	0	0	0	0	–	–	–	–
Tokelau	0	0	0	0	0	0	0	0	0	0	0	0
Tonga	0	2	0	2	0	0	0	0	0	0	0	0
Tuvalu	0	0	0	0	0	0	0	0	0	0	0	0
Vanuatu	0	1	0	1	0	1	0	1	0	0	0	0
Wallis and Futuna Islands	0	1	0	1	0	0	0	0	0	0	0	0
Total Oceania	**4 316**	**489**	**1 539**	**3 266**	**2 715**	**350**	**709**	**2 356**	**4 165**	**2 081**	**1 410**	**4 836**
Bangladesh	9	25	0	34	65	30	0	95	58	245	0	303
Bhutan	32	0	23	9	0	0	0	0	0	1	1	0
India	2 554	277	72	2 758	4 048	507	5	4 550	4 183	1 427	309	5 301
Maldives	0	4	0	4	0	0	0	0	0	1	0	1
Nepal	30	2	0	32	15	4	2	17	13	7	1	19
Pakistan	481	275	0	756	372	94	0	466	1 010	303	0	1 313
Sri Lanka	22	60	25	56	21	2	0	23	25	146	1	170
Total South Asia	**3 127**	**642**	**121**	**3 648**	**4 521**	**637**	**7**	**5 152**	**5 289**	**2 130**	**312**	**7 106**
Brunei Darussalam	0	6	0	6	0	0	0	0	0	4	1	3
Cambodia	7	4	5	5	0	0	0	0	0	28	0	28
Indonesia	5 376	244	3 600	2 020	5 587	681	2 761	3 507	7 223	327	3 510	4 040
Lao People's Democratic Republic	24	1	5	20	0	0	0	0	0	3	0	3
Malaysia	7 767	370	7 208	929	124	272	14	382	941	2 469	243	3 167
Myanmar	113	4	53	64	40	1	0	41	45	39	0	84
Philippines	418	272	59	631	212	77	23	266	1 097	618	145	1 571
Singapore	355	314	147	522	0	90	86	4	87	699	163	623
Thailand	3 000	247	2 758	488	1 146	375	179	1 343	3 796	656	1 088	3 363

Country/area	Wood-based panels (1 000 m³)				Pulp for paper (1 000 m³)				Paper and paperboard (1 000 m³)			
	Production	Imports	Exports	Consumption	Production	Imports	Exports	Consumption	Production	Imports	Exports	Consumption
Timor-Leste	0	0	0	0	0	0	0	0	0	0	0	0
Viet Nam	460	575	28	1 007	710	163	0	873	888	597	30	1 454
Total Southeast Asia	**17 520**	**2 036**	**13 864**	**5 692**	**7 818**	**1 659**	**3 062**	**6 415**	**14 077**	**5 441**	**5 179**	**14 338**
Total Asia and the Pacific	**98 081**	**16 730**	**25 379**	**89 432**	**45 500**	**15 502**	**4 102**	**56 900**	**122 107**	**21 043**	**17 207**	**125 942**
Belarus	895	190	359	726	66	26	0	92	285	141	86	340
Republic of Moldova	10	25	0	34	0	0	0	0	0	27	8	19
Russian Federation	8 962	1 512	2 359	8 115	6 882	60	1 780	5 162	7 434	1 221	2 701	5 954
Ukraine	1 662	662	426	1 898	0	92	1	91	791	738	164	1 365
Total CIS countries	**11 529**	**2 388**	**3 145**	**10 773**	**6 948**	**178**	**1 781**	**5 345**	**8 510**	**2 126**	**2 958**	**7 678**
Albania	37	112	0	149	0	4	0	4	3	18	1	20
Bosnia and Herzegovina	28	147	25	150	20	34	0	54	118	60	42	136
Bulgaria	389	235	381	243	135	15	48	102	326	246	87	485
Croatia	161	244	126	279	107	1	43	65	564	213	121	656
Czech Republic	1 566	650	984	1 233	766	171	346	591	1 042	1 249	769	1 523
Estonia	423	215	321	316	136	3	51	88	73	139	108	104
Hungary	720	410	375	755	19	164	1	182	553	729	435	847
Latvia	450	154	401	203	0	1	0	1	57	136	43	150
Lithuania	378	429	132	675	0	2	0	2	119	170	94	195
Montenegro	0	11	0	11	0	0	0	0	0	3	0	3
Poland	7 357	1 571	2 132	6 796	1 062	413	31	1 444	2 857	2 580	1 470	3 967
Romania	1 376	781	932	1 225	150	15	2	163	432	294	121	605
Serbia	91	352	39	404	14	11	1	24	59	98	1	156
Slovakia	981	507	363	1 125	626	90	92	624	888	403	771	520
Slovenia	495	349	291	553	112	197	25	284	760	267	561	466
The former Yugoslav Republic of Macedonia	0	88	3	86	0	1	0	1	20	57	8	69
Total Eastern Europe	**14 452**	**6 256**	**6 506**	**14 203**	**3 147**	**1 123**	**641**	**3 629**	**7 871**	**6 662**	**4 632**	**9 900**
Andorra	0	2	0	2	0	0	0	0	0	2	0	2
Austria	3 449	813	2 860	1 402	1 678	697	228	2 147	5 213	1 291	4 113	2 391
Belgium	2 585	1 972	3 089	1 468	509	808	913	404	1 897	3 957	3 298	2 556
Channel Islands	–	–	–	–	–	–	–	–	–	–	–	–
Denmark	345	1 622	161	1 806	0	72	0	72	423	1 208	308	1 323
Faeroe Islands	0	1	0	1	0	0	0	0	0	2	0	1
Finland	2 074	362	1 623	813	13 615	267	2 762	11 120	14 140	458	12 906	1 693
France	6 657	2 085	3 926	4 816	2 331	2 217	556	3 992	10 006	6 230	5 269	10 967
Germany	17 400	4 153	7 565	13 988	2 938	4 978	1 035	6 881	22 656	11 176	13 909	19 923
Gibraltar	0	0	0	0	0	0	0	0	0	0	0	0
Greece	860	445	136	1 169	0	76	1	75	510	1 118	68	1 560
Holy See	–	–	–	–	–	–	–	–	–	–	–	–

TABLE 5 (CONT.)

Production, trade and consumption of wood-based panels, pulp and paper, 2006

| Country/area | Wood-based panels | | | | Pulp for paper | | | | Paper and paperboard | | | |
| | (1 000 m³) | | | | (1 000 m³) | | | | (1 000 m³) | | | |
	Production	Imports	Exports	Consumption	Production	Imports	Exports	Consumption	Production	Imports	Exports	Consumption
Iceland	0	24	0	24	0	0	0	0	0	40	0	40
Ireland	937	382	827	492	0	4	1	3	45	510	71	484
Isle of Man	–	–	–	–	–	–	–	–	–	–	–	–
Italy	5 740	2 000	1 128	6 612	683	3 672	29	4 326	10 011	5 175	3 492	11 694
Liechtenstein	–	–	–	–	–	–	–	–	–	–	–	–
Luxembourg	450	43	357	136	0	0	0	0	0	150	31	119
Malta	0	31	0	31	0	0	0	0	0	34	0	34
Monaco	–	–	–	–	–	–	–	–	–	–	–	–
Netherlands	10	1 871	363	1 518	109	1 293	495	907	3 367	3 367	3 169	3 565
Norway	603	316	268	651	2 303	58	488	1 873	2 109	492	1 821	780
Portugal	1 306	381	943	744	2 065	67	1 038	1 094	1 644	736	1 297	1 083
San Marino	–	–	–	–	–	–	–	–	–	–	–	–
Spain	5 091	1 817	1 274	5 634	2 888	926	990	2 824	6 893	4 812	2 719	8 986
Sweden	842	1 093	905	1 029	12 066	445	3 163	9 348	12 066	1 008	10 849	2 225
Switzerland	964	617	903	678	165	533	31	666	1 685	1 157	1 304	1 538
United Kingdom	3 498	3 685	510	6 673	287	1 315	19	1 583	5 813	7 756	1 001	12 568
Total Western Europe	**52 811**	**23 714**	**26 838**	**49 687**	**41 636**	**17 428**	**11 749**	**47 316**	**98 478**	**50 679**	**65 624**	**83 534**
Total Europe	**78 792**	**32 359**	**36 488**	**74 663**	**51 732**	**18 729**	**14 171**	**56 289**	**114 859**	**59 468**	**73 214**	**101 112**
Anguilla	–	–	–	–	–	–	–	–	–	–	–	–
Antigua and Barbuda	0	4	0	4	0	0	0	0	0	0	0	0
Aruba	0	6	0	6	0	0	0	0	0	1	0	1
Bahamas	0	28	0	28	0	0	0	0	0	10	9	0
Barbados	0	30	0	30	0	1	0	1	2	9	0	11
Bermuda	–	–	–	–	–	–	–	–	–	–	–	–
British Virgin Islands	0	1	0	1	0	0	0	0	0	0	0	0
Cayman Islands	0	5	0	5	0	0	0	0	0	1	0	1
Cuba	149	15	0	164	1	3	0	4	27	61	1	87
Dominica	0	3	1	2	0	0	0	0	0	1	0	0
Dominican Republic	0	58	0	58	0	1	0	1	130	207	1	336
Grenada	0	4	0	4	0	0	0	0	0	0	0	0
Guadeloupe	0	23	0	23	0	0	0	0	0	6	0	6
Haiti	0	2	0	2	0	0	0	0	0	9	0	9
Jamaica	0	70	0	70	0	0	0	0	0	35	0	35
Martinique	0	7	0	7	0	0	0	0	0	5	0	5
Montserrat	0	0	0	0	0	0	0	0	0	0	0	0
Netherlands Antilles	0	5	0	4	0	0	0	0	0	5	2	3
Puerto Rico	–	–	–	–	–	–	–	–	–	–	–	–
Saint Kitts and Nevis	0	1	0	1	0	0	0	0	0	0	0	0
Saint Lucia	0	7	0	7	0	0	0	0	0	10	0	10
Saint Vincent and the Grenadines	0	2	0	2	0	0	0	0	0	3	0	3

Country/area	Wood-based panels (1 000 m³)				Pulp for paper (1 000 m³)				Paper and paperboard (1 000 m³)			
	Production	Imports	Exports	Consumption	Production	Imports	Exports	Consumption	Production	Imports	Exports	Consumption
Trinidad and Tobago	0	44	0	44	0	4	0	4	0	100	1	99
Turks and Caicos Islands	0	1	0	1	0	0	0	0	0	0	0	0
United States Virgin Islands	–	–	–	–	–	–	–	–	–	–	–	–
Total Caribbean	**149**	**318**	**2**	**465**	**1**	**10**	**0**	**11**	**159**	**464**	**15**	**607**
Belize	0	4	1	3	0	2	1	1	0	2	1	1
Costa Rica	65	50	33	82	10	33	0	42	20	392	22	390
El Salvador	0	30	0	29	0	1	1	1	56	141	7	189
Guatemala	31	55	16	70	0	3	0	3	31	301	18	314
Honduras	14	25	7	32	7	0	0	7	95	156	3	248
Nicaragua	8	10	5	13	0	0	0	0	0	30	0	30
Panama	7	27	0	34	0	2	0	2	0	98	28	70
Total Central America	**125**	**200**	**62**	**263**	**17**	**41**	**2**	**56**	**202**	**1 121**	**79**	**1 243**
Argentina	1 322	55	622	756	937	91	212	816	2 080	727	208	2 599
Bolivia	30	7	19	18	0	0	0	0	0	50	0	50
Brazil	9 121	432	2 812	6 741	11 271	379	6 217	5 433	8 518	931	1 820	7 629
Chile	2 285	144	1 369	1 059	3 484	14	2 822	676	1 231	401	563	1 069
Colombia	245	174	38	381	387	164	1	550	990	511	170	1 331
Ecuador	261	67	121	207	2	24	0	26	100	232	21	311
Falkland Islands	0	0	0	0	0	0	0	0	0	0	0	0
French Guiana	0	3	0	3	0	0	0	0	0	0	0	0
Guyana	34	4	35	3	0	0	0	0	0	6	0	5
Paraguay	161	5	31	135	0	0	0	0	13	75	3	85
Peru	65	125	40	151	17	68	0	85	102	320	15	406
South Georgia and the South Sandwich Islands	–	–	–	–	–	–	–	–	–	–	–	–
Suriname	1	7	2	6	0	0	0	0	0	4	0	4
Uruguay	7	40	5	42	34	8	0	42	98	89	40	147
Venezuela (Bolivarian Republic of)	695	67	61	701	148	116	0	264	693	270	9	954
Total South America	**14 228**	**1 131**	**5 154**	**10 205**	**16 280**	**864**	**9 252**	**7 892**	**13 825**	**3 614**	**2 848**	**14 590**
Total Latin America and the Caribbean	**14 501**	**1 649**	**5 218**	**10 933**	**16 298**	**914**	**9 254**	**7 958**	**14 186**	**5 198**	**2 943**	**16 441**
Canada	17 633	2 534	13 017	7 150	23 481	313	10 727	13 067	18 189	2 895	14 200	6 884
Greenland	0	5	0	5	0	0	0	0	0	1	0	1
Mexico	259	1 965	237	1 988	314	1 206	20	1 500	4 844	2 997	292	7 548
Saint Pierre and Miquelon	0	1	0	0	0	0	0	0	0	0	0	0
United States of America	44 359	20 401	2 189	62 571	53 074	6 285	5 771	53 588	84 317	16 524	9 644	91 196
Total North America	**62 251**	**24 906**	**15 442**	**71 714**	**76 869**	**7 804**	**16 518**	**68 155**	**107 350**	**22 416**	**24 137**	**105 629**

TABLE 5 (CONT.)
Production, trade and consumption of wood-based panels, pulp and paper, 2006

Country/area	Wood-based panels (1 000 m³)				Pulp for paper (1 000 m³)				Paper and paperboard (1 000 m³)			
	Production	Imports	Exports	Consumption	Production	Imports	Exports	Consumption	Production	Imports	Exports	Consumption
Armenia	1	64	0	65	0	0	0	0	4	12	0	17
Azerbaijan	0	230	1	229	0	0	0	0	3	35	3	35
Georgia	10	5	0	15	0	0	0	0	0	6	0	6
Kazakhstan	10	503	1	512	0	1	0	1	81	143	16	209
Kyrgyzstan	0	34	0	34	0	0	0	0	2	17	0	19
Tajikistan	0	0	0	0	0	0	0	0	0	0	0	0
Turkmenistan	0	3	1	2	0	0	0	0	0	1	0	1
Uzbekistan	0	257	3	254	0	3	3	0	11	61	5	67
Total Central Asia	**21**	**1 097**	**7**	**1 111**	**0**	**4**	**3**	**1**	**102**	**275**	**24**	**353**
Afghanistan	1	12	0	13	0	0	0	0	0	1	0	1
Bahrain	0	55	1	54	0	12	0	12	15	26	18	23
Cyprus	3	129	0	132	0	2	0	2	0	104	0	104
Iran (Islamic Republic of)	677	350	7	1 020	507	75	0	582	411	571	4	977
Iraq	5	99	0	104	11	0	0	11	33	13	0	46
Israel	181	289	13	456	15	139	17	137	275	553	20	808
Jordan	0	169	19	149	8	76	0	84	54	154	32	176
Kuwait	0	154	0	154	0	9	0	9	56	126	27	155
Lebanon	46	304	2	348	0	35	0	35	103	170	13	260
Occupied Palestinian Territory	–	–	–	–	–	–	–	–	–	–	–	–
Oman	0	136	0	135	0	1	0	1	0	66	4	62
Qatar	0	125	0	125	0	5	0	5	0	26	15	11
Saudi Arabia	0	267	0	267	0	64	0	64	279	774	26	1 027
Syrian Arab Republic	27	353	1	379	0	50	0	50	75	196	2	269
Turkey	4 989	896	561	5 324	138	475	2	611	1 643	2 068	175	3 536
United Arab Emirates	0	418	26	392	0	18	0	18	81	480	52	509
Yemen	0	133	0	133	0	0	0	0	0	82	0	82
Total Western Asia	**5 929**	**3 888**	**630**	**9 187**	**679**	**961**	**19**	**1 622**	**3 025**	**5 410**	**389**	**8 046**
Total Western and Central Asia	**5 950**	**4 985**	**637**	**10 298**	**679**	**965**	**22**	**1 622**	**3 127**	**5 685**	**413**	**8 399**
TOTAL WORLD	**262 092**	**81 929**	**84 107**	**259 914**	**194 668**	**44 715**	**45 343**	**194 040**	**364 579**	**116 468**	**118 283**	**362 764**

SOURCE: FAOSTAT (ForesSTAT), last accessed 28 August 2008.

TABLE 6

Forestry sector's contribution to employment and gross domestic product, 2006

Country/area	Employment					Gross value added				
	Roundwood production	Wood processing	Pulp and paper	Total for the forestry sector		Roundwood production	Wood processing	Pulp and paper	Total for the forestry sector	
	(1 000)	(1 000)	(1 000)	(1 000)	(% of total labour force)	(US$ million)	(US$ million)	(US$ million)	(US$ million)	(% contribution to GDP)
Burundi	0	2	0	2	0.0	10	5	0	15	1.8
Cameroon	12	8	1	20	0.3	236	74	13	324	1.9
Central African Republic	2	2	0	4	0.2	133	10	1	144	11.1
Chad	1	0	–	1	0.0	122	0	–	122	1.9
Congo	4	3	0	7	0.5	45	27	–	72	1.1
Democratic Republic of the Congo	6	0	–	6	0.0	185	2	–	186	2.3
Equatorial Guinea	1	0	–	1	0.5	86	2	–	87	0.9
Gabon	8	4	0	12	1.9	171	118	0	290	3.0
Rwanda	1	1	–	1	0.0	30	1	–	31	1.3
Saint Helena	–	–	–	–	–	–	–	–	–	–
Sao Tome and Principe	–	–	–	–	–	–	–	–	–	–
Total Central Africa	**35**	**19**	**1**	**55**	**0.1**	**1 017**	**239**	**15**	**1 271**	**2.0**
British Indian Ocean Territory	–	–	–	–	–	–	–	–	–	–
Comoros	–	–	–	–	–	18	–	–	18	4.4
Djibouti	–	–	–	–	–	0	–	–	0	0.1
Eritrea	0	0	0	0	0.0	0	0	0	1	0.1
Ethiopia	1	2	2	5	0.0	630	4	9	643	5.2
Kenya	1	10	8	19	0.1	242	20	106	368	1.7
Madagascar	2	41	1	44	0.4	148	8	0	157	3.1
Mauritius	1	1	1	2	0.4	7	4	12	23	0.4
Mayotte	–	–	–	–	–	–	–	–	–	–
Réunion	0	0	0	0	0.1	2	8	8	18	0.1
Seychelles	–	–	–	–	–	0	–	–	0	0.1
Somalia	0	1	–	1	0.0	15	1	–	15	0.6
Uganda	2	1	1	4	0.0	354	16	9	379	4.0
United Republic of Tanzania	3	6	6	15	0.1	205	1	22	228	1.9
Total East Africa	**11**	**61**	**19**	**90**	**0.1**	**1 623**	**62**	**166**	**1 851**	**2.1**
Algeria	0	11	2	13	0.1	37	118	66	220	0.2
Egypt	1	3	18	21	0.1	131	7	157	296	0.3
Libyan Arab Jamahiriya	0	1	0	2	0.1	57	4	2	62	0.1
Mauritania	0	0	0	0	0.0	1	0	–	1	0.1
Morocco	13	8	5	26	0.2	343	80	126	549	0.9
Sudan	1	2	1	4	0.0	57	15	36	107	0.3
Tunisia	4	9	4	16	0.4	106	147	149	402	1.4
Western Sahara	–	–	–	–	–	–	–	–	–	–
Total Northern Africa	**19**	**34**	**30**	**83**	**0.1**	**731**	**372**	**535**	**1 638**	**0.4**

TABLE 6 (CONT.)
Forestry sector's contribution to employment and gross domestic product, 2006

Country/area	Employment					Gross value added				
	Roundwood production	Wood processing	Pulp and paper	Total for the forestry sector		Roundwood production	Wood processing	Pulp and paper	Total for the forestry sector	
	(1 000)	(1 000)	(1 000)	(1 000)	(% of total labour force)	(US$ million)	(US$ million)	(US$ million)	(US$ million)	(% contribution to GDP)
Angola	2	1	0	3	0.0	260	2	1	262	0.6
Botswana	0	0	0	1	0.1	25	1	5	30	0.4
Lesotho	1	0	–	1	0.1	67	–	–	67	5.0
Malawi	1	1	0	2	0.0	40	2	8	50	2.6
Mozambique	12	3	0	15	0.1	221	2	2	224	3.1
Namibia	0	0	0	0	0.1	–	6	0	6	0.1
South Africa	45	37	34	116	0.5	920	948	1 677	3 545	1.6
Swaziland	1	2	3	6	1.5	11	10	60	80	5.2
Zambia	1	1	2	5	0.1	547	61	21	629	5.9
Zimbabwe	1	6	7	13	0.2	49	14	12	74	5.3
Total Southern Africa	**63**	**51**	**47**	**161**	**0.3**	**2 139**	**1 044**	**1 785**	**4 969**	**1.6**
Benin	1	0	–	1	0.0	103	5	0	108	2.6
Burkina Faso	2	2	0	4	0.1	88	0	–	88	1.5
Cape Verde	0	1	–	1	0.5	20	0	–	20	2.0
Côte d'Ivoire	19	8	1	28	0.4	672	96	33	801	5.0
Gambia	0	1	–	1	0.1	1	0	–	1	0.2
Ghana	12	30	1	43	0.4	542	202	10	754	7.2
Guinea	9	1	–	10	0.2	39	6	–	45	1.7
Guinea-Bissau	1	0	–	1	0.1	18	2	–	20	6.3
Liberia	1	1	–	2	0.1	113	9	–	121	17.7
Mali	1	0	–	1	0.0	102	0	–	102	1.9
Niger	1	0	–	1	0.0	98	0	7	105	3.3
Nigeria	24	3	18	45	0.1	1 506	32	282	1 819	1.4
Senegal	1	0	1	2	0.0	65	3	9	77	0.9
Sierra Leone	0	0	0	1	0.0	84	0	0	85	4.8
Togo	1	0	–	1	0.0	31	2	–	33	1.6
Total West Africa	**73**	**46**	**20**	**140**	**0.1**	**3 480**	**357**	**342**	**4 179**	**2.2**
Total Africa	**202**	**211**	**117**	**530**	**0.1**	**8 991**	**2 075**	**2 843**	**13 908**	**1.3**
China	1 172	937	1 409	3 518	0.4	13 687	8 834	18 687	41 208	1.3
Democratic People's Republic of Korea	19	4	4	26	0.2	220	33	46	299	2.5
Japan	32	150	211	393	0.6	892	9 590	22 422	32 904	0.7
Mongolia	1	1	0	1	0.1	2	3	1	7	0.2
Republic of Korea	12	25	63	99	0.4	1 498	1 099	5 877	8 473	1.1
Total East Asia	**1 235**	**1 115**	**1 686**	**4 037**	**0.4**	**16 298**	**19 559**	**47 033**	**82 890**	**1.0**
American Samoa	–	–	–	–	–	–	–	–	–	–
Australia	11	42	21	74	0.7	695	2 806	2 061	5 562	0.8

Country/area	Employment					Gross value added				
	Roundwood production	Wood processing	Pulp and paper	Total for the forestry sector		Roundwood production	Wood processing	Pulp and paper	Total for the forestry sector	
	(1 000)	(1 000)	(1 000)	(1 000)	(% of total labour force)	(US$ million)	(US$ million)	(US$ million)	(US$ million)	(% contribution to GDP)
Cook Islands	–	–	–	–	–	–	–	–	–	–
Fiji	0	2	1	3	0.6	29	52	11	92	3.4
French Polynesia	0	0	0	0	0.3	–	–	–	–	–
Guam	0	–	–	0	0.0	–	–	–	–	–
Kiribati	–	–	–	–	–	0	–	–	0	0.0
Marshall Islands	–	–	–	–	–	–	–	–	–	–
Micronesia (Federated States of)	–	–	–	–	–	–	–	–	–	–
Nauru	–	–	–	–	–	–	–	–	–	–
New Caledonia	0	0	0	0	0.1	1	1	–	2	0.0
New Zealand	7	16	5	28	1.4	691	897	584	2 172	2.1
Niue	–	–	–	–	–	–	–	–	–	–
Northern Mariana Islands	–	–	–	–	–	–	–	–	–	–
Palau	–	–	–	–	–	–	–	–	–	–
Papua New Guinea	8	4	–	12	0.4	316	84	–	400	6.7
Pitcairn Islands	–	–	–	–	–	–	–	–	–	–
Samoa	0	0	–	1	0.8	6	8	–	14	3.2
Solomon Islands	8	0	–	8	3.0	53	4	–	57	16.7
Tokelau	–	–	–	–	–	–	–	–	–	–
Tonga	0	0	0	0	0.3	1	0	0	1	0.5
Tuvalu	–	–	–	–	–	–	–	–	–	–
Vanuatu	0	1	–	1	1.4	3	10	–	13	3.5
Wallis and Futuna Islands	–	–	–	–	–	–	–	–	–	–
Total Oceania	**36**	**65**	**27**	**128**	**0.8**	**1 794**	**3 862**	**2 657**	**8 313**	**1.0**
Bangladesh	1	11	24	36	0.0	997	76	45	1 118	1.7
Bhutan	1	2	–	3	0.2	49	12	–	61	6.9
India	246	55	180	481	0.1	5 927	132	1 092	7 151	0.9
Maldives	–	0	–	0	0.0	–	–	–	–	–
Nepal	12	4	3	19	0.1	318	5	8	330	4.3
Pakistan	30	5	22	58	0.1	288	9	213	510	0.4
Sri Lanka	17	4	3	23	0.3	199	17	31	247	1.0
Total South Asia	**308**	**80**	**231**	**619**	**0.1**	**7 777**	**251**	**1 388**	**9 416**	**0.9**
Brunei Darussalam	1	0	–	2	0.9	3	6	–	9	0.1
Cambodia	0	1	0	1	0.0	139	5	29	173	2.8
Indonesia	69	148	104	321	0.3	3 283	3 896	2 386	9 564	2.5
Lao People's Democratic Republic	1	2	0	3	0.1	103	1	0	104	3.0
Malaysia	88	126	35	248	2.3	2 423	1 514	661	4 598	3.0
Myanmar	24	21	3	48	0.2	35	1	1	38	0.3
Philippines	8	20	21	49	0.1	94	157	308	560	0.5

TABLE 6 (CONT.)
Forestry sector's contribution to employment and gross domestic product, 2006

Country/area	Employment					Gross value added				
	Roundwood production	Wood processing	Pulp and paper	Total for the forestry sector		Roundwood production	Wood processing	Pulp and paper	Total for the forestry sector	
	(1 000)	(1 000)	(1 000)	(1 000)	(% of total labour force)	(US$ million)	(US$ million)	(US$ million)	(US$ million)	(% contribution to GDP)
Singapore	0	2	4	6	0.3	–	38	181	218	0.2
Thailand	8	62	67	137	0.4	149	333	1 211	1 693	0.8
Timor-Leste	–	–	–	–	–	1	–	–	1	0.4
Viet Nam	22	120	70	212	0.5	674	370	328	1 372	2.4
Total Southeast Asia	**221**	**502**	**304**	**1 027**	**0.4**	**6 904**	**6 322**	**5 105**	**18 331**	**1.7**
Total Asia and the Pacific	**1 800**	**1 763**	**2 248**	**5 811**	**0.3**	**32 774**	**29 994**	**56 183**	**118 951**	**1.0**
Belarus	33	46	23	103	1.9	180	399	97	677	2.1
Republic of Moldova	4	1	2	6	0.3	7	10	5	21	0.7
Russian Federation	383	336	131	849	1.1	1 029	3 381	2 417	6 828	0.8
Ukraine	152	60	23	235	0.9	427	350	326	1 103	1.2
Total CIS countries	**572**	**443**	**178**	**1 193**	**1.1**	**1 643**	**4 141**	**2 845**	**8 628**	**0.9**
Albania	2	1	0	2	0.1	6	4	3	13	0.2
Bosnia and Herzegovina	7	5	2	14	0.7	129	85	17	232	2.5
Bulgaria	15	23	11	49	1.2	59	97	77	232	0.9
Croatia	9	12	5	26	1.2	115	186	161	462	1.3
Czech Republic	35	83	20	138	2.5	832	1 225	596	2 654	2.1
Estonia	7	19	2	28	3.6	148	345	43	536	3.7
Hungary	8	37	16	61	1.4	142	319	330	790	0.8
Latvia	29	34	1	65	5.0	232	353	26	610	3.4
Lithuania	9	25	2	35	1.8	121	449	70	641	2.4
Montenegro	1	2	0	3	1.1	14	10	0	25	1.3
Poland	49	138	42	229	1.1	965	2 003	1 386	4 353	1.5
Romania	57	77	17	151	1.4	435	1 116	318	1 869	1.7
Serbia	6	11	9	26	0.7	81	39	72	191	0.6
Slovakia	12	34	7	54	1.8	221	470	266	957	1.9
Slovenia	6	11	5	22	2.3	125	263	181	569	1.8
The former Yugoslav Republic of Macedonia	4	3	1	8	0.8	18	3	3	24	0.4
Total Eastern Europe	**257**	**515**	**141**	**912**	**1.4**	**3 643**	**6 966**	**3 548**	**14 158**	**1.6**
Andorra	–	0	0	0	1.0	–	–	–	–	–
Austria	7	36	17	61	1.5	1 494	2 661	2 013	6 168	2.1
Belgium	2	14	14	31	0.7	191	1 114	1 424	2 729	0.8
Channel Islands	–	–	–	–	–	–	–	–	–	–
Denmark	4	15	7	25	0.9	201	1 002	602	1 805	0.8
Faeroe Islands	–	–	–	–	–	–	–	–	–	–
Finland	23	32	35	90	3.6	3 329	1 918	5 082	10 329	5.7

Country/area	Employment					Gross value added				
	Roundwood production	Wood processing	Pulp and paper	Total for the forestry sector		Roundwood production	Wood processing	Pulp and paper	Total for the forestry sector	
	(1 000)	(1 000)	(1 000)	(1 000)	(% of total labour force)	(US$ million)	(US$ million)	(US$ million)	(US$ million)	(% contribution to GDP)
France	31	87	74	191	0.7	5 107	4 147	5 653	14 907	0.7
Germany	44	165	134	342	0.8	2 259	9 315	12 324	23 898	0.9
Gibraltar	–	–	–	–	–	–	–	–	–	–
Greece	5	25	8	37	0.8	116	428	328	872	0.3
Holy See	–	–	–	–	–	–	–	–	–	–
Iceland	0	1	0	1	0.6	1	33	7	40	0.3
Ireland	2	9	3	15	0.9	132	524	278	934	0.5
Isle of Man	–	–	–	–	–	–	–	–	–	–
Italy	41	171	66	278	1.1	940	6 778	5 547	13 265	0.8
Liechtenstein	0	1	0	1	3.6	1	–	–	1	0.0
Luxembourg	0	1	0	1	0.5	12	64	38	115	0.3
Malta	–	0	0	0	0.2	0	3	5	8	0.2
Monaco	–	0	–	0	0.2	–	–	–	–	–
Netherlands	2	17	22	41	0.6	65	1 341	1 873	3 279	0.6
Norway	5	15	7	26	1.1	274	1 245	716	2 234	0.8
Portugal	12	57	12	81	1.6	809	1 022	923	2 755	1.7
San Marino	–	0	0	0	1.5	–	–	–	–	–
Spain	23	100	51	174	1.0	1 252	3 770	4 252	9 273	0.8
Sweden	22	38	36	95	2.0	3 108	2 706	6 939	12 753	3.8
Switzerland	5	35	12	52	1.3	311	2 537	1 316	4 164	1.1
United Kingdom	11	86	69	166	0.6	246	4 839	4 633	9 719	0.4
Total Western Europe	**239**	**904**	**567**	**1 709**	**0.9**	**19 848**	**45 447**	**53 955**	**119 249**	**0.9**
Total Europe	**1 067**	**1 861**	**886**	**3 815**	**1.1**	**25 134**	**56 554**	**60 348**	**142 036**	**1.0**
Anguilla	–	–	–	–	–	0	–	–	0	0.0
Antigua and Barbuda	–	–	–	–	–	–	–	–	–	–
Aruba	–	0	–	0	0.1	–	–	–	–	–
Bahamas	0	0	0	0	0.1	0	0	3	3	0.0
Barbados	0	0	1	2	1.2	0	8	40	49	1.8
Bermuda	–	0	0	0	0.1	0	–	–	0	0.0
British Virgin Islands	–	–	–	–	–	0	–	–	0	0.0
Cayman Islands	–	–	–	–	–	–	–	–	–	–
Cuba	10	24	1	36	0.6	17	94	2	113	0.2
Dominica	–	–	–	–	–	1	–	–	1	0.5
Dominican Republic	0	0	1	1	0.0	7	–	9	17	0.1
Grenada	0	0	0	0	0.1	1	–	–	1	0.2
Guadeloupe	–	–	–	–	–	0	0	–	0	0.0
Haiti	1	0	0	1	0.0	5	0	–	6	0.1
Jamaica	1	1	1	3	0.2	6	2	52	60	0.6
Martinique	0	–	–	0	0.0	0	0	–	0	0.0
Montserrat	–	–	–	–	–	–	–	–	–	–
Netherlands Antilles	–	–	0	0	0.2	–	–	–	–	–

TABLE 6 (CONT.)
Forestry sector's contribution to employment and gross domestic product, 2006

Country/area	Employment					Gross value added				
	Roundwood production	Wood processing	Pulp and paper	Total for the forestry sector		Roundwood production	Wood processing	Pulp and paper	Total for the forestry sector	
	(1 000)	(1 000)	(1 000)	(1 000)	(% of total labour force)	(US$ million)	(US$ million)	(US$ million)	(US$ million)	(% contribution to GDP)
Puerto Rico	–	1	2	3	0.2	–	50	62	112	0.1
Saint Kitts and Nevis	–	–	–	–	–	0	–	–	0	0.0
Saint Lucia	–	–	–	–	–	0	–	4	4	0.5
Saint Vincent and the Grenadines	–	–	0	0	0.2	2	–	–	2	0.5
Trinidad and Tobago	1	2	2	5	0.8	16	10	42	68	0.4
Turks and Caicos Islands	–	–	–	–	–	–	–	–	–	–
United States Virgin Islands	–	0	0	0	0.1	–	–	–	–	–
Total Caribbean	**14**	**29**	**9**	**52**	**0.3**	**57**	**165**	**215**	**436**	**0.2**
Belize	1	2	0	3	2.6	7	11	1	19	1.7
Costa Rica	1	7	5	13	0.7	12	42	118	171	0.8
El Salvador	4	5	4	13	0.4	121	2	70	193	1.1
Guatemala	7	1	2	10	0.2	483	51	52	587	2.0
Honduras	3	15	2	20	0.7	73	49	27	149	1.8
Nicaragua	3	1	–	4	0.2	40	45	7	92	1.9
Panama	1	1	2	3	0.2	26	6	36	67	0.4
Total Central America	**20**	**32**	**13**	**65**	**0.4**	**762**	**206**	**311**	**1 279**	**1.3**
Argentina	54	32	30	116	0.7	311	156	1 098	1 564	0.8
Bolivia	4	3	2	9	0.2	92	111	38	241	2.7
Brazil	306	503	201	1 010	1.2	18 198	3 953	6 055	28 206	2.8
Chile	44	27	15	86	1.2	448	1 008	2 153	3 609	2.6
Colombia	3	4	18	25	0.1	140	166	503	810	0.7
Ecuador	13	4	7	24	0.4	277	427	190	893	2.3
Falkland Islands	–	–	–	–	–	–	–	–	–	–
French Guiana	0	0	–	0	0.3	2	2	–	4	0.1
Guyana	3	5	–	8	1.9	18	13	–	31	4.1
Paraguay	3	2	1	5	0.2	163	81	56	301	3.6
Peru	19	6	6	31	0.3	278	204	458	940	1.1
South Georgia and the South Sandwich Islands	–	–	–	–	–	–	–	–	–	–
Suriname	1	3	0	4	2.2	6	9	–	15	0.9
Uruguay	4	3	2	8	0.8	163	35	40	239	1.2
Venezuela (Bolivarian Republic of)	8	25	33	66	0.5	540	629	484	1 653	1.0
Total South America	**463**	**616**	**314**	**1 393**	**0.8**	**20 638**	**6 793**	**11 074**	**38 506**	**2.1**
Total Latin America and the Caribbean	**497**	**677**	**337**	**1 510**	**0.7**	**21 457**	**7 164**	**11 600**	**40 221**	**1.9**

Country/area	Employment					Gross value added				
	Roundwood production	Wood processing	Pulp and paper	Total for the forestry sector		Roundwood production	Wood processing	Pulp and paper	Total for the forestry sector	
	(1 000)	(1 000)	(1 000)	(1 000)	(% of total labour force)	(US$ million)	(US$ million)	(US$ million)	(US$ million)	(% contribution to GDP)
Canada	63	128	84	275	1.6	7 229	13 488	11 284	32 000	2.7
Greenland	–	–	–	–	–	–	–	–	–	–
Mexico	84	85	125	293	0.6	1 720	1 855	3 477	7 052	0.9
Saint Pierre and Miquelon	0	–	–	0	0.0	–	–	–	–	–
United States of America	85	565	459	1 109	0.7	18 528	37 400	52 500	108 428	0.8
Total North America	**232**	**778**	**667**	**1 677**	**0.8**	**27 477**	**52 743**	**67 261**	**147 480**	**1.0**
Armenia	2	1	0	3	0.2	4	1	2	7	0.1
Azerbaijan	2	2	0	4	0.1	2	3	1	6	0.0
Georgia	6	3	0	9	0.3	11	4	1	16	0.2
Kazakhstan	10	1	3	14	0.2	29	13	17	59	0.1
Kyrgyzstan	3	1	1	5	0.2	2	1	1	4	0.2
Tajikistan	2	0	0	3	0.1	0	0	0	1	0.0
Turkmenistan	2	0	–	2	0.1	0	0	–	0	0.0
Uzbekistan	6	1	0	7	0.1	2	9	2	14	0.1
Total Central Asia	**34**	**8**	**5**	**47**	**0.1**	**51**	**32**	**24**	**107**	**0.1**
Afghanistan	–	–	–	–	–	4	2	–	5	0.1
Bahrain	–	0	0	0	0.1	–	1	6	6	0.0
Cyprus	1	2	1	3	0.8	3	91	30	123	0.8
Iran (Islamic Republic of)	7	8	22	36	0.1	270	86	355	711	0.3
Iraq	–	0	6	6	0.1	–	12	26	39	0.1
Israel	1	5	8	14	0.5	–	121	312	433	0.3
Jordan	0	4	4	8	0.3	–	16	70	86	0.7
Kuwait	–	1	1	2	0.2	–	26	56	82	0.1
Lebanon	–	3	6	10	0.7	1	63	189	253	1.1
Occupied Palestinian Territory	–	1	0	2	0.7	–	12	9	21	0.6
Oman	–	1	1	2	0.2	–	20	15	35	0.1
Qatar	–	5	0	5	1.5	–	73	16	89	0.2
Saudi Arabia	1	21	13	35	0.4	–	–	279	279	0.1
Syrian Arab Republic	1	16	2	19	0.3	4	87	31	122	0.4
Turkey	33	89	45	167	0.5	1 342	609	834	2 786	0.7
United Arab Emirates	–	1	4	5	0.4	–	–	81	81	0.0
Yemen	–	3	2	5	0.1	–	31	22	54	0.3
Total Western Asia	**44**	**160**	**115**	**318**	**0.3**	**1 624**	**1 250**	**2 331**	**5 205**	**0.3**
Total Western and Central Asia	**78**	**168**	**120**	**365**	**0.2**	**1 675**	**1 282**	**2 355**	**5 312**	**0.3**
TOTAL WORLD	**3 876**	**5 459**	**4 374**	**13 709**	**0.4**	**117 508**	**149 811**	**200 589**	**467 908**	**1.0**

SOURCE: FAO 2008b.

References

ACTED. 2006. *Eco-tourism in Tajikistan: key challenges and opportunities*. Paris, Agency for Technical Cooperation and Development (available at www.untj.org/principals/minutes/TourismACTED.ppt).

Alexander, S.J., Weigand, J. & Blatner, K.A. 2002. Nontimber forest product commerce. *In* E.T. Jones, R.J. McLain & J. Weigand, eds. *Non timber forest products in the United States*. Lawrence, USA, University of Kansas Press.

Amir, S. & Rechtman, O. 2006. *The development of forest policy in Israel in the 20th century: implications for the future*. Haifa, Israel, Center for Urban and Regional Studies, Technion – Israel Institute of Technology.

ARC. 2006. *RecFacts general statistics*. Washington, DC, American Recreation Coalition (available at www.funoutdoors.com/research).

ARC. 2007. *Forest service chief Gail Kimbell seeks to close gap between kids and nature*. News release. Washington, DC (available at www.funoutdoors.com/node/view/1933).

Asia Forest Network. 2008. *Where is the future for cultures and forests? Indigenous peoples and forest management in 2020*. Thematic study for the Asia-Pacific Forestry Sector Outlook Study. Bangkok, FAO Regional Office for Asia and the Pacific. (in press)

Audley, J.J., Papademetriou, D.G, Polaski, S. & Vaughan, S. 2004. *NAFTA's promise and reality: lessons from Mexico for the hemisphere*. Washington, DC, Carnegie Endowment for International Peace (available at www.carnegieendowment.org/files/nafta1.pdf).

Baudin, A., Eliasson, L., Gustafsson, A., Hagström, L., Helstad, K., Nyrud, A.Q., Sande, J.B., Haartveit, E.Y. & Ziethén, R. 2005. ICT and the wood industry. *In* L. Hetemäki & S. Nilsson, eds. *Information technology and the forest sector*, pp. 129–149. Vienna, International Union of Forest Research Organizations (IUFRO).

Becker, G., Coleman, E., Hetsch, S., Kazemi, Y. & Prins, K. 2007. *Mobilizing wood resources: can Europe's forests satisfy the increasing demand for raw material and energy under sustainable forest management*. Background paper, UNECE/FAO Workshop on Mobilizing Wood Resources. 11–12 January 2007. Geneva, Switzerland, United Nations Economic Commission for Europe (UNECE).

Beecher, J.F. 2007. Wood, trees and nanotechnology. *Nature Nanotechnology*, 2(8): 466–467 (available at www.nature.com/naturenanotechnology).

Bell, S., Tyrväinen, L., Sievänen, T., Pröbstl, U. & Simpson, M. 2007. Outdoor recreation and nature tourism: a European perspective. *Living Reviews in Landscape Research*, 1(2) (available at landscaperesearch.livingreviews.org/Articles/lrlr-2007-2/).

Bowe, S.A., Smith, R.L., Kline, D.E. & Araman, P.A. 2002. A segmental analysis of current and future scanning and optimizing technology in the hardwood sawmill industry. *Forest Products Journal*, 52(3): 68–76.

Brown, S. 2008. Beetle tree kill releases more carbon than fires. *Nature News*, 23 April (available at www.nature.com/news/2008/080423/full/news.2008.771.html).

Canadian Council of Forest Ministers. 2006. *Criteria and indicators of sustainable forest management in Canada: national status 2005*. Ottawa, Natural Resources Canada.

Carle, J.B. & Holmgren, P. 2008. Wood from planted forests – a global outlook 2005–2030. *Forest Products Journal*. (in press)

CEI-Bois, CEPF & CEPI. 2005. *Innovative and sustainable use of forest resources: Vision 2030*. A technology platform initiative by the European forest-based sector. Brussels, European Confederation of Woodworking Industries, Confederation of European Forest Owners & Confederation of European Paper Industries.

CIFOR. 2004. *Operationalising the ecosystem approach – re-inventing research*. Forest Livelihoods Briefs No. 2. Bogor, Indonesia, Center for International Forestry Research (available at www.cifor.cgiar.org/publications/pdf_files/livebrief/livebrief0402e.pdf).

CIFOR. 2008a. *Best Brazil nut practice in Bolivia*. Bogor, Indonesia (available at www.cifor.cgiar.org/Publications/Corporate/NewsOnline/NewsOnline43/brazil_nut.htm).

CIFOR. 2008b. *CIFOR's strategy, 2008–2018: Making a difference for forests and people*. Bogor, Indonesia (available at www.cifor.cgiar.org/publications/pdf_files/Books/CIFORStrategy0801.pdf).

Clark, M. 2007. PEFC presentation. Third International Workshop on Conformity Assessment, Rio de Janeiro, Brazil, 10–11 December 2007 (available at www.inmetro.gov.br/noticias/eventos/avaliacaoConformidade/Palestras/michael_clark.pdf).

Comisión Nacional Forestal, Mexico. 2008. North America Forest Outlook Study: Mexico country report. (unpublished SOFO 2009 contribution)

Conservation International. 2005. *Biodiversity hotspots*. Arlington, USA (available at www.biodiversityhotspots.org/xp/Hotspots/hotspots_by_region/).

Contreras-Hermosilla, A., Gregersen, H.M. & White, A. 2008. *Forest governance in countries with federal systems of government: lessons for decentralization.* Governance Brief No. 39. Bogor, Indonesia, CIFOR.

de Brito Cruz, C.H. & de Mello, L. 2006. *Boosting innovation performance in Brazil.* Economics Department Working Paper No. 532. ECO/WKP(2006)60. Paris, Organisation for Economic Co-operation and Development (OECD).

Dillaha, T., Ferraro P., Huang M., Southgate D., Upadhyaya, S. & Wunder, S. 2007. *Payments for watershed services: regional syntheses.* USAID PES Brief No. 7. Washington, DC, United States Agency for International Development (USAID) (available at www.cifor.cgiar.org/pes/_ref/publications/index.htm).

Eckelmann, C.M. 2005. *An overview of silvicultural practices in the Caribbean – historic development, current practices and emerging issues.* Bridgetown, FAO Subregional Office for the Caribbean. (unpublished)

EEA. 2005. *European environment outlook.* Copenhagen, European Environment Agency.

EEA. 2007. *Europe's environment: the fourth assessment.* Copenhagen.

Environment News Service. 2008a. Sugar for biofuel to displace Kenya's Tana Delta wildlife. 26 June (available at www.ens-newswire.com/ens/jun2008/2008-06-26-03.asp).

Environment News Service. 2008b. Florida to buy out sugar land for Everglades restoration. 25 June (available at www.ens-newswire.com/ens/jun2008/2008-06-25-01.asp).

European Commission. 2007. *Key figures 2007: towards a European research area – science, technology and innovation.* Brussels (available at cordis.europa.eu/documents/documentlibrary/97946551EN6.pdf).

Evans, J. & Turnbull, J. 2004. *Plantation forestry in the tropics.* 3rd edition. Oxford, UK, Oxford University Press.

FAO. 2001. *Global Forest Resources Assessment 2000. Main report.* FAO Forestry Paper No. 140. Rome (also available at www.fao.org/docrep/004/y1997e/y1997e00.htm).

FAO. 2003a. *Forestry Outlook Study for Africa: regional report – opportunities and challenges towards 2020.* FAO Forestry Paper No. 141. Rome (also available at www.fao.org/docrep/005/y4521e/y4521e00.htm).

FAO. 2003b. *Past trends and future prospects for the utilisation of wood for energy,* by J. Broadhead, J. Bahdon & A. Whiteman. Global Forest Products Outlook Study Working Paper GFPOS/WP/05. Rome.

FAO. 2004. *Will buying tropical forest carbon benefit the poor? Evidence from Costa Rica,* by S. Kerr, L. Lipper, A.S.P. Pfaff, R. Cavatassi, B. Davis, J. Hendy & A. Sanchez. ESA Working Paper No. 04-20. Rome (also available at ftp://ftp.fao.org/docrep/fao/007/ae402e/ae402e00.pdf).

FAO. 2005a. *In search of excellence: exemplary forest management in Asia and the Pacific,* by P.B. Durst, C. Brown, H.D. Tacio & M. Ishikawa, eds. RAP Publication 2005/2. Bangkok, FAO Regional Office for Asia and the Pacific (also available at www.fao.org/docrep/007/ae542e/ae542e00.htm).

FAO. 2005b. *State of the World's Forests 2005.* Rome (also available at www.fao.org/docrep/007/y5574e/y5574e00.htm).

FAO. 2005c. *Urban and peri-urban forestry and greening in West and Central Asia: experience, constraints and prospects,* by U. Akerlund. FOWECA Thematic Study Report. Rome (available at ftp://ftp.fao.org/docrep/fao/009/ah238e/ah238e00.pdf).

FAO. 2005d. *Wildlife issues and development prospects in West and Central Asia,* by R. Czudek. Wildlife Management Working Paper No. 9. Rome (also available at www.fao.org/docrep/010/ai548e/ai548e00.htm).

FAO. 2006a. *Global Forest Resources Assessment 2005 – progress towards sustainable forest management.* FAO Forestry Paper No. 147. Rome (also available at www.fao.org/docrep/008/a0400e/a0400e00.htm).

FAO. 2006b. *Global planted forests thematic study: results and analysis,* by A. Del Lungo, J. Ball & J. Carle. Planted Forests and Trees Working Paper No. 38. Rome (also available at www.fao.org/forestry/site/10368/en).

FAO. 2006c. *Tendencias y perspectivas del sector forestal en América Latina y el Caribe.* FAO Forestry Paper No. 145. Rome (also available at www.fao.org/docrep/009/a0470s/a0470s00.htm).

FAO. 2006d. *Global Forest Resources Assessment 2005 – report on fires in the Central Asian region and adjacent countries,* by J.G. Goldammer. Fire Management Working Paper FM/16. Rome (also available at www.fao.org/docrep/009/j7572e/j7572e00.htm).

FAO. 2006e. *Non wood forest products in Central Asia and Caucasus.* FOWECA Thematic Study. Rome (also available at www.fao.org/docrep/010/ag268e/ag268e00.htm).

FAO. 2006f. *Responsible management of planted forests: voluntary guidelines.* Planted Forests and Trees Working Paper No. 37. Rome (also available at www.fao.org/docrep/009/j9256e/j9256e00.htm).

FAO. 2007a. Female entrepreneurs in the NWFP world: shea butter sales change African women's plight. *Non-Wood News*, 15: 18 (available at www.fao.org/docrep/010/a1189e/a1189e00.htm).

FAO. 2007b. *FAO Statistical Yearbook 2005–2006*. Rome.

FAO. 2007c. *People, forests and trees in West and Central Asia: outlook for 2020*. FAO Forestry Paper No. 152. Rome (also available at www.fao.org/docrep/009/a0981e/a0981e00.htm).

FAO. 2007d. *The Global Environmental Facility and payments for ecosystem services: a review of current initiatives and recommendations for future PES support by GEF and FAO programs*, by P. Gutman & S. Davidson. PESAL Papers Series No. 1. Rome (also available at www.fao.org/es/esa/PESAL/attachments/PESAL1_Gutman.pdf).

FAO. 2007e. *The State of Food and Agriculture 2007: paying farmers for environmental services*. Rome (also available at www.fao.org/docrep/010/a1200e/a1200e00.htm).

FAO. 2007f. *Corporate private sector dimensions in planted forest investments*, by D.A. Neilson. Planted Forests and Trees Working Paper FP/40E (available at www.fao.org/forestry/site/10368/en/).

FAO. 2007g. *World bamboo resources: a thematic study prepared in the framework of the Global Forest Resources Assessment 2005*, by M. Lobovikov, S. Paudel, M. Piazza, H. Ren and J. Wu. Non-Wood Forest Products No. 18. Rome (also available at www.fao.org/docrep/010/a1243e/a1243e00.htm).

FAO. 2007h. *The role of coastal forests in the mitigation of tsunami impacts*, by K. Forbes & J. Broadhead. RAP Publication 2007/1. Bangkok, FAO Regional Office for Asia and the Pacific.

FAO. 2008a. ForeSTAT statistical database (available at faostat.fao.org).

FAO. 2008b. *Contribution of the forestry sector to national economies, 1990–2006*, by A. Lebedys. Rome. (in press)

FAO. 2008c. *Global forest product projections*, by R. Jonsson & A. Whiteman. Rome. (in press)

FAO. 2008d. *Forests and energy. Key issues*. FAO Forestry Paper No. 154. Rome (also available at www.fao.org/docrep/010/i0139e/i0139e00.htm).

FAO. 2008e. *Human wildlife conflict in Africa – causes, consequences and management strategies*. FAO Forestry Paper. Rome. (in press)

FAO. 2008f. *The status and trends of forests and forestry in West Asia*, by Q. Ma. Subregional report of the Forestry Outlook Study for West and Central Asia. Forestry Policy and Institutions Working Paper 20. Rome (also available at www.fao.org/docrep/010/k1652e/k1652e00.htm).

FAO. 2008g. *Forests and forestry in Central Asia and the Caucasus*, by M. Uemoto. Forest Policy and Institutions Working Paper. Rome. (in press)

FAO. 2008h. *Re-inventing forestry agencies – experiences of institutional restructuring in Asia and the Pacific*, by P. Durst, C. Brown, J. Broadhead, R. Suzuki, R. Leslie & A. Inoguchi, eds. RAP Publication 2008/05. Bangkok, FAO Regional Office for Asia and the Pacific.

FECOFUN. 2006. *About us*. Kathmandu, Federation of Community Forest Users Nepal (available at www.fecofun.org/about.php).

Friday Offcuts. 2008. Sovereign wealth Funds start investing in timberlands. 18 April (available at www.fridayoffcuts.com/dsp_newsletter.cfm?id=266).

Frost, P. & Bond, I. 2008. The CAMPFIRE programme in Zimbabwe: payments for wildlife services. *Ecological Economics*, 65(4): 776–787.

FSC. 2008. *Global FSC certificates: type and distribution.* Presentation (available at www.fsc.org/ppt_graphs.html).

Gorte, R.W. & Ramseur, J.L. 2008. *Forest carbon markets: potentials and drawbacks*, CRS Report for Congress, RL 34560. Washington, DC, Congressional Research Service.

Government of Cyprus. 2005. FOWECA country outlook paper for Cyprus. Nicosia, Forestry Department.

Government of Oman. 2005. FOWECA country outlook paper. Salalah, Oman, General Directorate of Animal Wealth, Ministry of Agriculture and Fisheries.

Griffin, C. 2007. *An engaged and engaging tourism safety and security policy dialogue*. Presented at the Association of Caribbean States (ASC) Regional Policy Dialogue on Tourist Safety and Security, St. Ann's, Trinidad and Tobago, 7 July (available at www.acs-aec.org/Tourism/TSS/english.htm).

Hamilton, K., Sjardin, M., Marcello, T. & Xu, G. 2008. *Forging a frontier: state of the voluntary carbon markets 2008*. Washington, DC, and London, Ecosystem Market Place and New Carbon Finance.

Hetemäki, L. & Nilsson, S. 2005. *Information technology and the forest sector*. IUFRO World Series Vol. 18. Vienna, IUFRO.

Houllier, F., Novotny, J., Päivinen, R., Rosén, K., Scarascia-Mugnozza, G. & von Teuffel, K. 2005. *Future forest research strategy for a knowledge-based forest cluster: an asset for sustainable Europe*. A vision paper of European national forest research institutes. EFI Discussion Paper 11. Joensuu, Finland, European Forest Research Institute.

IAASTD. 2008. *Executive summary of the synthesis report.* International Assessment of Agricultural Knowledge, Science and Technology for Development (available at www.agassessment.org/index.cfm?Page=IAASTD%20Reports&ItemID=2713).

IEA. 2007. *World Energy Outlook 2007.* Paris, International Energy Agency.

ILO. 2001. *Globalization and sustainability: the forest and wood industries on the move.* Geneva, Switzerland, International Labour Organization.

IMF. 2008. *World Economic Outlook April 2008.* World Economic and Financial Surveys. Washington, DC, International Monetary Fund.

Ince, P., Schuler, A., Spelter, H. & Luppold, W. 2007. *Globalization and structural change in the US forestry sector: an evolving context for sustainable forest management.* General Technical Report FPL-GTR-170. Washington, DC, USDA Forest Service.

International *Eucalyptus* **Genome Network (EUCAGEN).** 2007. Eucalyptus *tapped as the next tree genome to be sequenced, characterized & harnessed for bioenergy, carbon sequestration, and other industrial applications.* Press release (available at www.ieugc.up.ac.za).

IPCC. 2007. *Climate change 2007: synthesis report. Contribution of Working Groups I, II and III to the Fourth Assessment Report of the Intergovernmental Panel on Climate Change.* Geneva, Switzerland, Intergovernmental Panel on Climate Change (also available at www.ipcc.ch/ipccreports/ar4-syr.htm).

ITTO. 2005. *Achieving the ITTO Objective 2000 and sustainable forest management in Mexico: executive summary.* Report submitted to the International Tropical Timber Council by the Diagnostic Mission, ITTC XXXIX/5, thirty-ninth Session. Yokohama, Japan.

ITTO. 2006. *Status of tropical forest management 2005.* ITTO Technical Series No. 24. Yokohama, Japan.

ITTO. 2008. *Developing forest certification: towards increasing comparability and acceptance of forest certification systems worldwide.* ITTO Technical Series No. 29. Yokohama, Japan.

IUFRO. 2008. List of IUFRO's member organizations. Vienna, International Union of Forest Research Organizations (available at www.iufro.org/membership/members/).

Jenkins, M., Scherr, S.J. & Inbar, M. 2004. Markets for biodiversity services: potential roles and challenges. *Environment,* 46(4): 32–42.

Kaimowitz, D. 2007. *The prospects for reduced emissions from deforestation and degradation (REDD) in Mesoamerica.* New York, USA, Ford Foundation.

MacCleery, D. 1992. *American forests: a history of resiliency and recovery.* FS-540. Durham, USA, USDA – Forest Service.

MacCleery, D. 2008. Re-inventing the United States Forest Service: evolution from custodial management, to production forestry, to ecosystem management. *In: Re-inventing forestry agencies: experiences of institutional restructuring in Asia and the Pacific,* edited by P. Durst, C. Brown, J. Broadhead, R. Suzuki, R. Leslie & A. Inoguchi. RAP Publication 2008/05. Bangkok, FAO Regional Office for Asia and the Pacific.

Malagnoux, M., Sène, E.H. & Atzmon, N. 2007. Forests, trees and water in arid lands: a delicate balance. *Unasylva,* 229: 24–29.

Mantau, U., Steierer, F., Hetsch, S. & Prins, C. 2008. *Wood resources availability and demands – Part I: National and regional wood resource balances 2005.* Background paper to the UNECE/FAO Workshop on Wood Balances. Geneva, Switzerland, UNECE.

Martin, R.M. 2008. Deforestation, land-use change and REDD. *Unasylva,* 230: 3–11.

Metafore. 2007. *Green building programs* (available at www.metafore.org/index.php?p=Green_Building_Programs&s=176).

Mubin, S.F. 2004. *Outlook of the paper industry in the GCC* (available at www.highbeam.com/doc/1P3-777403821.html).

Muñoz-Piña, C., Guevara, A., Torres, J.M. & Braña, J. 2006. *Paying for the hydrological services of Mexico's forests.* Bogor, Indonesia, CIFOR.

Nair, C.T.S. 2004. What does the future hold for forestry education? *Unasylva,* 216: 3–9.

Natural Resources Canada. 2007a. *The state of Canada's forests. Annual report 2007.* Ottawa (also available at foretscanada.rncan.gc.ca/rpt).

Natural Resources Canada. 2007b. *Responding to the mountain pine beetle infestation* (available at canadaforests.nrcan.gc.ca/articletopic/138).

Natural Resources Canada. 2008a. North American Forest Outlook Study: Canada country report. Ottawa. (unpublished SOFO 2009 contribution)

Natural Resources Canada. 2008b. *Leading by innovation: forest science and technology, part 3* (available at canadaforests.nrcan.gc.ca/articletopic/83?format=print).

Neilson, D. 2007. Prospects for change in international investment patterns in forestry. Paper presented at the International Conference on the Future of Forests in Asia and the Pacific: Outlook for 2020, Chiang Mai, Thailand, 16–18 October 2007.

Nyrud, A.Q. & Devine, Å. 2005. E-Commerce. *In* L. Hetemäki & S. Nilsson, eds. *Information technology and the forest sector*, pp. 49–64. Vienna, IUFRO.

O'Loughlin, C. 2008. Institutional restructuring, reforms and other changes within the New Zealand forestry sector since 1986. *In: Re-inventing forestry agencies: Experiences of institutional restructuring in Asia and the Pacific*, edited by P. Durst, C. Brown, J. Broadhead, R. Suzuki, R. Leslie & A. Inoguchi. RAP Publication 2008/05. Bangkok, FAO Regional Office for Asia and the Pacific.

Parrotta, J.A. & Agnoletti, M. 2007. Traditional forest knowledge: challenges and opportunities. *Forest Ecology and Management*, 249: 1–4.

PATA. 2008. *Asia Pacific tourism revenues set to soar to US 4.6 trillion by 2010*. Pacific Asia Travel Association (available at www.forimmediaterelease.net/pm/1244.html).

PEFC. 2008. *Statistical figures on certification* (available at register.pefc.cz/statistics.asp).

Peksa-Blanchard, M., Dolzan, P., Grassi, A., Heinimo, J., Junginger, M., Ranta, T. & Walter, A. 2007. *Global wood pellets markets and industry: policy drivers, market status and raw material potential*. IEA Bioenergy Task 40 (available at www.bioenergytrade.org).

PwC. 2007a. *Risks and rewards: forest, paper & packaging in South America*. New York, USA, PricewaterhouseCoopers (also available at www.pwc.com).

PwC. 2007b. *South America becomes a global player in the forest, paper and packaging sector*. New York, USA (available at www.pwc.com).

Reitzer, R. 2007. *Technology roadmap: applications of nanotechnology in the paper industry* (available at www.jyu.fi/science/muut_yksikot/nsc/en/pdf/nanopap).

Renz, L. & Atienza, J. 2006. *International grantmaking update: a snapshot of US foundation trends*. New York, USA, Foundation Center (also available at foundationcenter.org/gainknowledge/research/pdf/intl_update_2006.pdf).

Roughley, D.J. 2005. *Nanotechnology: implications for the wood products industry*. Final report. North Vancouver, Canada, Forintek Canada Corporation.

Sample, V.A. 2007. Introduction to the 2007 Pinchot Distinguished Lecture (The rise and fall of the timber investment management organizations: ownership changes in US forestlands, by C.S. Binkley) (available at www.pinchot.org/files/Binkley.DistinguishedLecture.2007.pdf).

Schmitt, C.B., Belokurov, A., Besançon, C., Boisrobert, L., Burgess, N.D., Campbell, A., Coad, L., Fish, L., Gliddon, D., Humphries, K., Kapos, V., Loucks, C., Lysenko, I., Miles, L., Mills, C., Minnemeyer, S., Pistorius, T., Ravilious, C., Steininger, M. & Winkel, G. 2008. *Global ecological forest classification and forest protected area gap analysis – analyses and recommendations in view of the 10% target for forest protection under the Convention on Biological Diversity (CBD)*. Freiburg, Germany, Freiburg University Press.

Schulze, M., Grogan, J. & Vidal, E. 2007. Technical challenges to sustainable forest management in concessions on public lands in the Brazilian Amazon. *Journal of Sustainable Forestry*, 26(1): 61–76.

Shackleton, S.E., Shanley, P. & Ndoye, O. 2007. Invisible but viable: recognising local markets for non-timber forest products. *International Forestry Review*, 9(3): 697–712.

Sheppard, S.R.J. & Meitner, M.J. 2005. Using multi-criteria analysis and visualization for sustainable forest management planning with stakeholder groups. *Forest Ecology and Management*, 207: 171–187.

SME Toolkit India. 2008. *Environment Law Notification No: S.O. 525(E) (23-Apr-04) CAMPA – Constitution* (available at india.smetoolkit.org/india).

Task Force on the Future of American Innovation. 2005. *The knowledge economy: is the United States losing its competitive edge?* Washington, DC.

Temu, A. 2004. Africa south of the Sahara. *In* Trends in forestry education in Southeast Asia and Africa, 1993 to 2002: preliminary results of two surveys. *Unasylva*, 216: 17–21.

TerrAfrica. 2006. Background information on TerrAfrica. Brochure (available at www.terrafrica.org/default.asp?pid=7665368).

TIES. 2007. *Resources: ecotourism in Asia Pacific*. Washington, DC, The International Ecotourism Society.

TNC. 2004. *Final report: conservation easement working group*. Arlington, USA, The Nature Conservancy.

Tomaselli, I. & Sarre, A. 2005. Brazil gets new forest law. *ITTO Tropical Forest Update*, 15(4): 7.

UN. 2006a. *World economic and social survey 2006: diverging growth and development*. New York, USA, United Nations.

UN. 2006b. *Delivering as one*. Report of the Secretary-General's High-Level Panel. New York, USA.

UN. 2008a. World Urbanization Prospects: The 2007 Revision Population Database (available at esa.un.org/unup).

UN. 2008b. Common database (available at unstats.un.org/unsd/cdb/cdb_help/cdb_quick_start.asp).

UN. 2008c. Millennium Development Goals Indicators database series: terrestrial areas protected (available at mdgs.un.org/unsd/mdg/SeriesDetail.aspx?srid=783&crid).

UN. 2008d. *World Urbanization Prospects: The 2007 Revision – Highlights.* New York, USA (also available at www.un.org/esa/population/publications/wup2007/2007WUP_Highlights_web.pdf).

UN. 2008e. Commodity trade statistics database (available at comtrade.un.org).

UN. 2008f. Energy statistics database: charcoal data (available at unstats.un.org/unsd/energy/edbase.htm).

UNECE & FAO. 2005. *European Forest Sector Outlook Study 1960–2000–2020: main report.* Geneva, Switzerland.

UNECE & FAO. 2006a. *Proceedings, UNECE/FAO policy forum: public procurement policies on wood and paper products and their impacts on sustainable forest management and timber markets. Geneva, Switzerland, 5 October.* Rome (also available at www.fao.org/docrep/009/a0914e/a0914e00.htm).

UNECE & FAO. 2006b. *Forest Products Annual Market Review, 2005–2006.* Geneva Timber and Forest Study Paper 21. New York, USA, and Geneva, Switzerland, United Nations Publications (also available at www.unece.org/trade/timber/tc-publ.htm).

UNECE & FAO. 2007. *Forest Products Annual Market Review, 2006–2007.* Geneva Timber and Forest Study Paper 22. New York, USA, and Geneva, Switzerland, United Nations Publications (also available at www.unece.org/trade/timber/tc-publ.htm).

UNECE & FAO. 2008. *Forests and water. Note by the secretariat.* 66th session of the Timber Committee and 34th session of the European Forestry Commission, Rome, 21–24 October 2008. Geneva, Switzerland.

UNECE, FAO & ILO. 2003. *Report on the seminar on close to nature forestry.* Document TIM/EFC/WP.1/SEM.57/2003/3. Geneva, Switzerland, UNECE.

UNECE, MCPFE & FAO. 2007. *State of Europe's forests 2007 – The MCPFE report on sustainable forest management in Europe.* Warsaw, Ministerial Conference on the Protection of Forests in Europe (MCPFE) Liaison Unit.

UNEP. 2007. *Global environment outlook (GEO 4).* Nairobi, United Nations Environment Programme (also available at www.unep.org/geo/).

UNESCAP. 2007. *Millennium Development Goals: progress in Asia and the Pacific 2007.* Bangkok, UN Economic and Social Commission for Asia and the Pacific.

UNFF. 2004. *Traditional forest-related knowledge. Report of the Secretary-General.* E/CN.18/2004/7. United Nations Forum on Forests (available at www.un.org/esa/forests/documents-unff.html#4).

UNU. 2007. *Overcoming one of the greatest environmental challenges of our times: re-thinking policies to cope with desertification.* Policy brief based on the Joint International Conference "Desertification and International Policy Imperatives", Algiers, 17–19 December 2006. Tokyo, United Nations University.

UNWTO. 2008. *Asia Pacific Newsletter*, 11(1). Madrid, UN World Tourism Organization (also available at www.unwto.org/asia/news/en/newsle.php?op=2&subop=2).

US DoE. 2006. *Forest products industry technology roadmap*, prepared by Agenda 2020 Technology Alliance. Washington, DC, United States Department of Energy (also available at www.agenda2020.org).

US EPA. 2008. *Mitigation banking factsheet.* Washington, DC, United States Environmental Protection Agency (also available at www.epa.gov/OWOW/wetlands/facts/fact16.html).

US Forest Service. 2008. North American Forest Outlook Study: US country report. (unpublished SOFO 2009 contribution)

USAID. 2006. *Biodiversity in Latin America and the Caribbean.* Washington, DC, United States Agency for International Development (also available at www.usaid.gov/locations/latin_america_caribbean/issues/biodiversity_issue.html).

USAID. 2008. *The shea value chain: a uniquely African industry* (available at www.watradehub.com/index.php?option=com_content&task=view&id=507&Itemid=117).

USDA. 2004. *Rural poverty at a glance.* Rural Poverty Research Report No. 10. Washington, DC, United States Department of Agriculture.

USGBC. 2008. *What is LEED?* Washington, DC, United States Green Building Council (also available at www.usgbc.org/DisplayPage.aspx?CMSPageID=222).

van Ree, R. & Annevelink, B. 2007. *Status report biorefinery 2007.* Wageningen, the Netherlands, Agrotechnology and Food Sciences Group (also available at www.biorefinery.nl/publications).

Wang, T., Hamann, A., Aitken, S., O'Neill, G., Yanchuk, A. & Spittlehouse, D. 2008. Use of genetic variation in forest trees to adapt to changing climate. Presented at the conference "Adaptation of forests and forest management to changing climate with emphasis on forest health: a review of science, policies, and practices", Umeå, Sweden, 25–28 August.

WBCSD & WRI. 2007. *Sustainable procurement of wood and paper-based products.* Geneva, Switzerland & Washington, DC, World Business Council for Sustainable Development & World Resources Institute.

Welford, L. & Le Breton, G. 2008. Bridging the gap: PhytoTrade Africa's experience of the certification of natural products. *Forests, Trees and Livelihoods,* 18: 69–79.

World Bank. 2004. *Poverty in Mexico: an assessment of conditions, trends and government strategy,* Report No. 28612-ME. Washington, DC.

World Bank. 2006. *Doing business in 2007: how to reform.* Washington, DC.

World Bank. 2007a. *World Development Indicators.* Washington, DC.

World Bank. 2007b. *WDR/Latin America and the Caribbean: developed country subsidies, an obstacle for agricultural development. Agribusiness and biofuels are transforming the sector.* Series No. 2008/080/DEC (available at web. worldbank.org).

World Energy Council. 2005. *Regional energy integration in Africa.* London (also available at www.worldenergy.org/documents/integrationii.pdf).

World Resources Institute. 2007. *EarthTrends: January 2007 monthly update: forest certification and the path to sustainable forest management* (available at earthtrends.wri.org/updates/node/156).